Kevin Edelmann

Electroluminescence from Plasmonic Excitations in a Scanning Tunnelling Microscope

Experimental Condensed Matter Physics
Band 24

Herausgeber
Physikalisches Institut
Prof. Dr. David Hunger
Prof. Dr. Alexey Ustinov
Prof. Dr. Georg Weiß
Prof. Dr. Wolfgang Wernsdorfer
Prof. Dr. Wulf Wulfhekel

Eine Übersicht aller bisher in dieser Schriftenreihe erschienenen Bände finden Sie am Ende des Buchs.

Electroluminescence from Plasmonic Excitations in a Scanning Tunnelling Microscope

by
Kevin Edelmann

KIT Scientific Publishing

Karlsruher Institut für Technologie
Physikalisches Institut

Electroluminescence from Plasmonic Excitations
in a Scanning Tunnelling Microscope

Zur Erlangung des akademischen Grades eines Doktors der Naturwissenschaften von der KIT-Fakultät für Physik des Karlsruher Instituts für Technologie (KIT) genehmigte Dissertation

von Dipl. Phys. Kevin Edelmann aus Stuttgart

Tag der mündlichen Prüfung: 9. November 2018
Referent: Prof. Dr. Wulf Wulfhekel
Korreferent: Prof. Dr. David Hunger

Impressum

KIT Scientific Publishing

Karlsruher Institut für Technologie (KIT)
KIT Scientific Publishing
Straße am Forum 2
D-76131 Karlsruhe

KIT Scientific Publishing is a registered trademark
of Karlsruhe Institute of Technology.
Reprint using the book cover is not allowed.

www.ksp.kit.edu

Print on Demand 2019 – Gedruckt auf FSC-zertifiziertem Papier

ISSN 2191-9925
ISBN 978-3-7315-0923-3
DOI 10.5445/KSP/1000093148

"After sleeping through a hundred million centuries
we have finally opened our eyes on a sumptuous planet,
sparkling with colour,
bountiful with life.

Within decades we must close our eyes again.
Isn't it a noble, an enlightened way of spending
our brief time in the sun, to work at understanding
the universe and how we have come to wake up in it?"

— *Richard Dawkins*, in "Unweaving the rainbow"

Contents

List of abbreviations

CAD	Computer-aided design
CCD	Charge-coupled device
DLW	Direct laser writing
FT	Fourier transform
HOMO	Highest occupied molecular orbital
JT	Joule-Thomson
LDOS	Local density of states
LUMO	Lowest unoccupied molecular orbital
NA	Numerical aperture
PEEK	Polyether ether ketone
PTFE	Polytetrafluoroethylene
SEM	Scanning electron microscope
SPP	Surface plasmon polariton
STM	Scanning tunnelling microscopy
STML	Scanning tunnelling microscope induced luminescence
STS	Scanning tunnelling spectroscopy
TMP	Turbomolecular pump
UHV	Ultra-high vacuum

1 Introduction

Shortly after its invention in 1982 by Binnig and Rohrer [1, 2], scanning tunnelling microscopy (STM) was used to solve a long-standing puzzle in surface science, namely the atomic structure of the reconstructed Si(111) surface [3]. Since then, STM has evolved into one of the most versatile and powerful techniques in the toolbox of surface scientists due to its capability to not just image the surface topography at atomic resolution but to also perform elastic [4] and inelastic [5, 6] spectroscopy with spatial resolution, or to contact nanostructures at specific sites [7]. Apart from clean metal and semiconductor surfaces studies, STM has also provided unprecedented insights into related systems like adatoms and clusters [8], superconductors [9], heteroepitaxial thin films [10], and adsorbed molecules [11]. By using magnetic tips it became furthermore possible to investigate magnetic properties of surfaces at the atomic scale [12]. The traditional applications of STM have soon been complemented by more creative and exotic uses such as surface modification [13], ultra-fast scanning [14], radio-frequency bias modulation [15], optical pump-probe measurements [16], STM-induced chemical reactions [17, 18], and STM-induced luminescence (STML), of which the latter is at the focus of this work.

Light emission from localised plasmonic excitations in tunnelling junctions had been discovered even before the advent of STM [19]. But when the research group at IBM Zurich finally combined STM with a light collection and detection setup in the late 1980s [20–22] and presented pioneering works just shortly after [23–30], the additional experimental access to the surface at the sub-nanoscale spawned an entire new subdiscipline. The experimental difficulties that arise when trying to collect light from the tunnelling junction, however, caused STML to remain rather exotic despite its interesting opportunities. To this day, only a small number of groups are active in this field with a total publication output of less than 200 studies within the past 30 years, including the recently published, comprehensive reviews by Zhang *et al.* [31] and Kuhnke *et al.* [32].

In the tunnelling regime, i.e., at moderate tip-sample separation around 1 nm and bias voltages of a few volts, two effects have been found to cause light emission on clean surfaces. One of them is charge recombination in semiconducting samples when minority carriers are injected by the tip. This was used to characterise the diffusion length of minority carriers [33] and to probe the spatial variation of the recombination efficiency due to variations of the dopant concentration [23]. Even atomic resolution in the induced light emission was observed because the hole injection efficiency on a GaAs surface varies within a lattice unit cell [34]. Further studies addressed the tuning of exciton confinement in quantum wires [35] and the emission properties of single carbon nano-tubes [36–38]. Also molecular films acting as an organic semiconductor [39–44] have gained interest lately, as they facilitate the observation of strongly bound excitons due to reduced screening effects.

The other STM-induced luminescence mechanism, which is the main issue of this thesis, is the radiative decay of gap plasmons, i.e., collective charge carrier oscillations of conduction electrons in the tip and the sample, which are excited by one-electron inelastic tunnelling [26, 28]. Due to the electric-field enhancement caused by the sharp tip, the plasmon mode is highly localised around the tunnelling junction and the electron-to-photon conversion efficiency is drastically increased compared to tunnelling across flat electrodes. For sharp tips on flat noble metal surfaces it is estimated that one in 10^3 to 10^4 tunnelling electrons produces a photon, depending on the tip material and geometry [20, 26, 45–47].

The application of plasmons for sensing devices beyond the diffraction limit and the prospect of a new, disruptive signal-processing technology with optical interfaces based on surface plasmons has entailed a great deal of basic research in this field, particularly regarding the manipulation of localised and propagating plasmon modes [48]. Particularly localised plasmonic resonances provide a unique blend of opportunities for both miniaturisation, as the optically active volume is confined to a fraction of the wavelength, and operation at high band-width. In contrast to other techniques such as scanning near-field optical microscopy [49], electron energy-loss spectroscopy [50] or photoemission electron microscopy [51], STML cannot map the electric field distribution of plasmon eigenmodes of sample features because, first, all the light from the junction is collected regardless of its exact origin. Second, the STM tip is often intrusive to intrinsic plasmonic resonances of the sample by forming a coupled gap cavity. In STML, one can rather investigate how the location where electrons are injected into the system affects the light emission, and how features on the surface or the exact position of the tip interact with the gap cavity. In return for this short-coming, STML provides unmatched spatial resolution in the sense that the STM tip can be positioned very precisely, and the optically active area can be highly confined to the tip apex, either because of the electric field confinement of the tip in the case of plasmons, or because of the short diffusion length of minority carriers in the case of semiconductors.

As a result, STML may not be the technique of choice for developing or characterising photonic circuits, but it has provided unique physical insights into the electromagnetic coupling of electrodes at the ultimate atomic scale. The coupling strength across the tunnelling gap and the frequency of the gap resonance have been demonstrated to be sensitive to changes of the tip-sample distance as little as 10 pm, and modifications of the gap cavity by vertical atomic steps [52–56] have been observed as well as by the even finer influence of the lateral atomic structure on the tip-sample distance at constant current [57]. Particularly the influence of atomic steps will be discussed in detail in Secs. 4.2.3 and 5.2.2 alongside measurements that have been performed in the scope of this thesis. In the literature, tip induced gap plasmons have been tuned by embedding silver particles in a dielectric matrix [58], by different protrusions of a granular surface [59], by the thickness of the metallic film [60], and by specifically designed resonators [61]. It was further shown that the light emission allows to distinguish between separated Ag and Cu clusters deposited on a glass substrate due to their distinctly different dielectric constant [62]. Also the size and shape of individual nano-clusters [63] were observed to alter the gap cavity, and by engineering the plasmon resonance with the shape of specific gold nano-clusters, even directional light emission was achieved [64]. The coupling between the tip and a nano-cluster can furthermore not only be controlled by the vertical but also by the lateral tip position, so that either vertical or lateral plasmon modes could be induced in an isolated silver cluster, depending on whether the tip was placed above or next to it [65, 66]. By using non-plasmonic tips made of tungsten and operating at large tip-sample separations, STML can also be performed in a less intrusive fashion, so that intrinsic Mie-resonances in individual nano-clusters were observed, which mainly involve the conduction electrons of the clusters [67, 68]. Apart from acting as a local light source, the STM tip can also be used to locally inject surface plasmon polaritons into a waveguide [69]. Regardless of the investigated system, any STML experiment additionally benefits from all other advantages that STM has to offer, i.e., the possibility to characterise the topography and electronic structure of the sample while simultaneously probing its optical response.

One major feature of STML is that the obtained light signal not only conveys information about the electromagnetic environment of the tip, but depends to a large extent on the tunnelling current as well. Changes of the current linearly affect the excitation rate, which can be exploited to monitor the tunnelling current on a picosecond scale due to the minimal delay between a current variation and a corresponding change of emission intensity, which is not possible with any other technique [70, 71]. A more subtle influence of the tunnelling current on the light emission is exerted by variations of the inelastic tunnelling probability, which can occur independently of the elastic tunnelling probability because different end states are involved. This effect is difficult to assess in a traditional STM measurement in which the inelastic fraction of the measured total current is small. STML, however, is solely a consequence of inelastic tunnelling and, therefore, highly sensitive to variations of the inelastic channel. Central findings of Ch. 5 will later be ascribed to this effect.

While the unique information that can be obtained from the light emission from clean metal surfaces is interesting in itself, STML also provides a unique opportunity for single-molecule studies, which are at the focus of most recent publications in this field. The problem how to chemically engineer molecules in order to achieve efficient fluorescence is long solved, for as long as the molecules are in free space or in solution. But applications like organic light-emitting diodes ultimately require to interface the optically active molecules with metallic electrodes. On metallic surfaces, however, the electronic states of the substrate and the molecular states hybridise so that electronic molecular excitations are efficiently dissipated into the substrate. This nonradiative damping causes the electronic levels of a molecule to broaden and fluorescence to be dramatically quenched [72, 73]. A common solution to this problem in organic light-emitting diodes is to separate the optically active area from the metallic leads by organic exciton-blocking and carrier injection layers [74]. This is not an ideal solution due to ohmic losses in the separating layers, and understanding the electron transport through the random molecular configurations particularly at the interfaces, as well as all excitation and de-excitation mechanisms in a system with that many degrees of freedom is a daunting task. As a consequence, optimisation of these devices is so far only possible in a heuristic fashion. In order to gain insights into the fundamental process of converting electrical energy into light inside a molecule, it is of advantage to reduce the investigated system to a single molecule. In such a system with greatly reduced degrees of freedom, well-defined and reproducible experimental conditions can be achieved.

Investigating STM induced luminescence in single-molecules is particularly promising as STM is a long-established tool for single-molecule studies. Therefore, the molecular adsorption and its electronic properties can be characterised within the same experiment, which offers valuable additional insights into the studied system on the single-molecule level, whereas many other characterisation techniques are limited to large ensembles of molecules. For these reasons, considerable effort has recently been put into STML studies of single molecules adsorbed on metal surfaces. Molecules can provide additional, potentially long-lived electronic states for charge recombination, or emit light by radiative de-excitation of electronically or vibrationally excited states [75]. Recent activities in the field have explored spatial and time resolution of vibrationally excited molecular states [76–78], charging effects [79], emission colour engineering [80], superirradiance and energy transfer due to inter-molecular interactions [81, 82], as well as interactions with the plasmonic cavity formed by the tip and the sample [83–85]. In order to contribute to this growing and interesting field, the main goal of this thesis is to develop and commission a new STM with an efficient light collection system, which is capable of single-molecule studies.

On the experimental side, however, the subject poses a major challenge. Both the tunnelling current and the conversion efficiency in the STM junction are low, so the light signal to be obtained is weak. Moreover, spectroscopic resolution is important to identify different luminescence mechanisms, especially if molecules are involved. For that purpose, the collected light has to be split into different wavelength bins, reducing the detectable signal per bin even further. It is therefore of paramount importance to collect as many of the emitted photons as possible, i.e., cover a large solid angle around the STM junction, whereas the demand for mechanical rigidity of the STM often leaves little space to realise such a collection system.

So far, mainly lenses, parabolic or ellipsoid mirrors, and optical fibres very close to the junction have been employed in STML setups. In the case of lenses and mirrors, the light from the junction is collimated *in-situ* in close proximity and refocussed *ex-situ* onto either a photon counter or a spectrograph. When using lenses and fibres, the covered solid angle is only a few percent of the half-space, and by combining several lenses around the junction, coverages of up to 27 % have been achieved [86]. By surrounding the STM junction with a large curved mirror, on the other hand, coverages of up to 65 % have been achieved [25].

However, the main issue with lenses and especially curved mirrors is that both the collimating and the focussing optics need to be finely adjusted for every tip. That requires complex and expensive adjustment components, both inside and outside the vacuum chamber — for each lens, if several of them are installed [86]. These optics and adjustment components close to the junction easily compromise the accessibility of the sample and the tip for *in-situ* transfers. In case of cryogenic STMs, also openings in the radiation shields are necessary to clear the light path, which introduces heating of the STM by radiation unless the optical pathway is carefully designed to require only pinhole openings [86]. Optical fibres do not have this disadvantage. However, they need to be extremely close to the tip and due to their limited acceptance angle they are only of limited use for directly collecting photons from a point source. Still, they need to point precisely to the junction, which requires *in-situ* adjustability as well.

The idea to combine a parabolic mirror with the STM tip and to take care of focussing the mirror during fabrication *ex-situ* was already explored by Khang *et al.* [87]. In this work, this idea is taken further and the tip and mirror are fabricated as one integral piece instead of an assembly, the mirror fully surrounds the junction and the optic is miniaturised such that the diameter of the collimated light beam is small enough to be directly coupled into an optical multimode fibre. Hereby, six problems are overcome simultaneously: (i) the collection optics covers the highest possible solid angle, (ii) the collection optics does not prevent access to the tip or the sample during transfers, (iii) no focussing on the tip is required, and the collimated output beam does not need to be refocussed, (iv) during operation, the tip and collection optics are always at the same, constant temperature and cannot drift with respect to each other, (v) the collection optics does not need to be perfect due to the lack of free-space routing, and (vi) there is no need for a direct optical pathway from ambient conditions through the cryogenic shields to the STM. For the fabrication process, 3D direct laser writing (DLW) based on two-photon polymerisation [88] is employed, as it is ideally suited for creating tailored free-form microoptics with the necessary precision.

To facilitate single-molecule studies, the STM was integrated in a low-temperature ultra-high vacuum (UHV) cryostat, which enables investigations of clean samples over extended periods of time without further contamination or surface diffusion.

In this thesis, the design and assembly of the experimental setup will be described, following by commissioning measurements on the well-known Au(111) surface in order to demonstrate the performance of the system. Finally, the newly developed STM was used to investigate the light emission from bilayer and trilayer Co nano-islands epitaxially grown on a Cu(111) surface. It was shown by Downes *et al.* [62] that it is possible to distinguish different metals in STML due to their different dielectric functions if they form large, separated clusters on an isolating substrate. The goal of the investigation presented in this thesis is to push the limits of this technique and test if such a distinction is also possible if only few atomic layers of a different metal are present in the STM junction, which are not isolated from the substrate.

2 Fundamentals

2.1 Scanning tunnelling microscopy

In STM, a precisely positionable, atomically sharp metallic needle — referred to as *tip* — is used to probe a conductive surface at distances around and below 1 nm by means of the tunnelling current which flows across the vacuum gap upon applying a bias voltage. Owing to the exponential distance dependence of quantum tunnelling, the vast majority of the tunnelling current flows between the top-most atom of the tip apex and an equally small area of the sample. As a result, STM provides access to a variety of local surface properties with unmatched resolution.

2.1.1 Theory

Quantum tunnelling has been a textbook example of quantum mechanics for decades and arises from the finite probability of a particle to pass through a thin potential barrier. This is a consequence of the Schrödinger equation, which is, regarding space, a second-order differential equation whose solutions must be continuous and continuously differentiable. Therefore, a particle's wavefunction cannot abruptly vanish at the onset of a potential barrier but rather extends into the barrier where it finally decays exponentially over distance. Provided the barrier is sufficiently thin, the wavefunction then continues on the other side with an accordingly reduced amplitude. This simple model can be applied to STM in the sense that conduction electrons can freely propagate in a metal but are restricted to the solid's boundaries by the work function. The vacuum gap between the sample and the tip, therefore, comprises a thin potential barrier created by the work functions on both sides. With this model one can already understand the exponential distance dependence of the tunnelling current.

A more adequate way to describe the tunnelling process also taking into account the electronic structure of both the tip and the sample is the transfer-Hamiltonian approach introduced by Bardeen [89]. Due to their separation, the interaction between

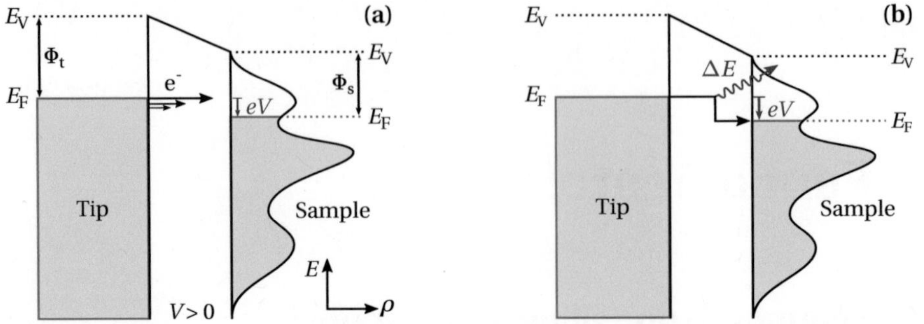

Figure 2.1: One-dimensional sketch of the energy landscape of tunnelling electrons in STM. The electronic density of states in the tip is assumed featureless for clarity. (a) Elastic tunnelling, (b) inelastic tunnelling at the voltage threshold $eV = \Delta E$.

the tip and the sample is considered weak, which justifies applying first-order time-dependent perturbation theory. At this, the transition rate of the electrons is according to Fermis golden rule determined by the overlap of the unperturbated eigenfunctions Ψ_μ and ψ_ν of the tip and sample, respectively, here denoted as the matrix element $M_{\mu\nu} = \langle \Psi_\mu | H_\mathrm{T} | \psi_\nu \rangle$ with H_T being the perturbation Hamiltonian. Apart from this, energy conservation is demanded, and that electrons transition from occupied to unoccupied states. The latter is ensured by weighing the contributions of all combinations of beginning and end states with the Fermi–Dirac function at their respective eigenenergies $E_{\mu,\nu}$. If a bias voltage V is applied to the sample, a net current can be obtained by summing over the contributions of all involved initial and end states:

$$I_T = \frac{2\pi e}{\hbar} \sum_{\nu,\mu} \big(f_\mathrm{t}(E_\mu) - f_\mathrm{s}(E_\nu + eV)\big) \left|M_{\mu\nu}\right|^2 \delta\big(E_\mu - E_\nu - eV\big), \tag{2.1}$$

whereas subscripts s and t denote quantities related to the sample and tip, respectively. In this notation, a positive bias polarity leads to electrons tunnelling from the tip to the sample. The difficulty is now finding an expression for $M_{\mu\nu}$, which Bardeen further showed to take the form

$$M_{\mu\nu} = \frac{\hbar^2}{2m} \int_S \mathrm{d}\mathscr{S} \cdot \left(\Psi^*_\mu \Delta \psi_\nu - \psi_\nu \Delta \Psi^*_\mu\right). \tag{2.2}$$

This surface integral is to be evaluated over any surface within the barrier region.

Bardeens work predates STM and was originally meant to describe metal-insulator-metal tunnelling junctions. Tersoff and Hamann later applied this method to STM [90], sparking other works based on the same idea with similar results [91–93]. In STM, the tunnelling current is carried for the very most part by the topmost atom of the tip. To model the tip it should therefore suffice to consider this topmost atom. Assuming low temperatures and that the eigenstates in the tip have the form of a spherically symmetric s-wave function allows to calculate Eq. 2.2. One finds that the tunnelling current depends on the local density of states (LDOS) ρ of the sample at the position of the tip defined as the centre $\vec{r}_0$ of its spherical wavefunction

$$I(\vec{r}_0, V, d) = \frac{e}{\hbar} \int_0^{eV} \rho_s(\vec{r}_0, E_{F,s} + \epsilon)\rho_t(E_{F,t} - eV + \epsilon)\, T(\epsilon, V, d)\, \mathrm{d}\epsilon. \tag{2.3}$$

The transfer probability T may now be simplified using a one-dimensional, semi-classical Wentzel-Kramers-Brillouin approximation as [94]

$$T(E, V, d) = \exp\left(-\frac{2d}{\hbar}\sqrt{m_e \cdot \left(\Phi_s + \Phi_t + eV - 2E\right)}\right), \tag{2.4}$$

which again involves the work functions $\Phi_{s,t}$ of the sample and the tip, and reproduces the exponential distance dependence already known from the one-dimensional solution of the Schrödinger equation. This also means that electrons at the Fermi edge contribute most to the tunnelling current, which is sketched in Fig. 2.1(a) together with the energy landscape of the tip and the sample in this model. A popular simplification for small bias voltages is to neglect the energy-dependence of the transfer probability. For convenience, the LDOS of the tip is also often considered featureless, such that the LDOS of the sample can be extracted from the first derivative of the tunnelling current with respect to the bias voltage

$$\frac{\mathrm{d}I}{\mathrm{d}V} \propto \rho_s(\vec{r}_0, E_{F,s} + eV). \tag{2.5}$$

This is commonly referred to as scanning tunnelling spectroscopy (STS). In practise, however, measurements of the differential conductance often bear the signature of the electronic structure of the tip, which can drastically change the experimental results [4]. In order to ascribe any observed feature in such measurements to the LDOS of the sample, one requires carefully prepared tips and needs to compare spectroscopic data acquired with different tips.

A tunnelling electron can also interact with local excitations, which is called inelastic tunnelling. For these processes, the energy conservation term in Eq. 2.1 is corrected for the energy which is transferred to the excitation, and the electron ends up in a final state of correspondingly lower energy. Thus, more end states become available to the tunnelling electrons if the tunnelling bias exceeds the necessary threshold for the excitation. This is the situation sketched in Fig. 2.1(b). Terms for the additional available end states then add to Eq. 2.1 and increase the tunnelling current. Experimentally, this can be observed as a kink in the current-voltage characteristics at the excitation threshold $e|V| = \Delta E$, which is independent from the voltage polarity. To recognise such a signature more easily, one usually consults the first or second derivative of the current-voltage dependence which then exhibit symmetric steps or a dip-peak pair around the origin, respectively. The list of inelastic excitations observed in STM so far includes phonons [5], molecular vibrations [6], spin flips [95, 96], magnons [97], and — essentially to this work — plasmons [26].

2.1.2 Experimental implementation

The high spatial resolution of STM depends on the sharpness of the tip and, more critically, the possibility to position the tip at picometre precision. The metallic tips are typically prepared by electrochemically etching a wire [98], but the exact preparation technique and achieved tip radius are not critical for the STM performance as long as the final apex is rigidly supported by the rest of the material, and is not hanging on a lose whisker. Tips that provided atomic resolution in STM have even been achieved by mechanically grinding a wire to a point or cutting a wire at an angle [2, 3, 99]. Positioning the tip, on the other hand, is achieved by utilising the piezoelectric effect. Below a critical temperature, ferroelectric ceramics exhibit domains of spontaneous electric polarisation because the centres of positive and negative charges within the crystal lattice are spatially separated in the ground state. Thus, an internal electric field is generated which is directly linked to the crystal structure, so that mechanical deformation results in a change of the electric field and — more important for STM — vice versa. By applying a moderate external voltage to a ferroelectric crystal whose domains have previously been aligned by a high voltage, the size of the crystal changes approximately linearly with the voltage. In STM, the tip is usually mounted on a piezo element in the shape of a thin-walled tube which carries out the tip movement in all three directions. This is preferred compared to an orthogonal arrangement of three different actuators because such a tube is smaller and more rigid. The piezo tube is equipped with an electrode on the inside and four segmented electrodes on the outside, each of which spans over a quarter of the outer diameter [100]. Applying differential voltages to opposite outside electrodes results in bending of the tube, which approximately translates to a lateral movement of the tube's end in a plane as the bending is small compared to the size of the tube. By varying the voltage applied to the inside electrode, finally, the length of the tube changes. Therefore, the position of the tip can be varied in both lateral as well as the vertical direction with high accuracy, but within a very limited range. During STM operation, one also needs to account for some non-ideal properties of the piezo elements, which are hysteresis and a specific time-dependence of the displacement response upon voltage changes. The latter is commonly referred to as

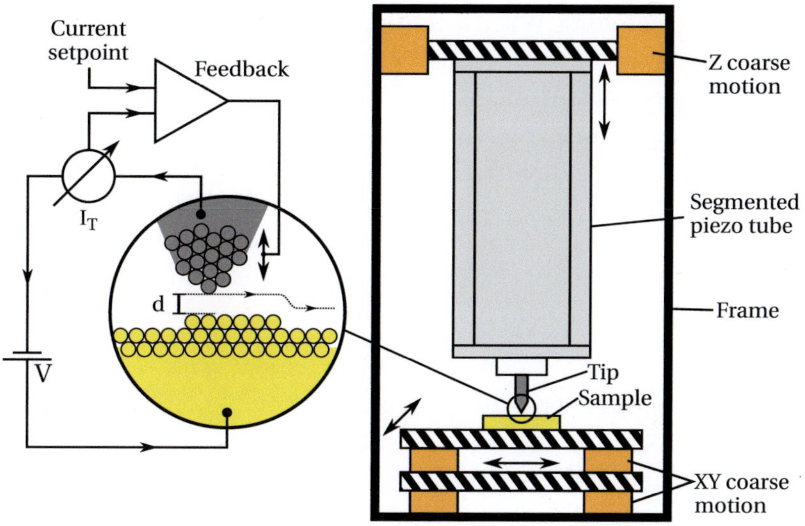

Figure 2.2: Working principle of an STM in the constant current mode, and sketch of a popular implementation according to [101] using a segmented piezo tube for positioning the tip on the pico- to nanometre scale and several coarse motion stages to approach the tip at a desired location on the sample.

piezo creep and can compromise the long-term stability of the tip position even hours after large displacements.

The basic blueprint of a popular implementation for STM is sketched in Fig. 2.2. Apart from the precision positioning of the tip using the piezo tube, another positioning system needs to be implemented in order to approach the assembly of piezo tube and tip in small steps to the sample, as especially the vertical range of the piezo tube is limited to a few hundred nanometres. In many designs, this coarse motion is based on the stick-slip principle [102], which will be discussed in detail in Sec. 3.2.2. Many designs also implement a lateral coarse motion for the sample to be able to roughly choose a spot on the sample to investigate. In Fig. 2.2, two stacked lateral coarse motion stages for moving the sample are shown because this corresponds to the design choice for the STM presented in this work. However, these are entirely optional and there are designs with only one stage [103] or no motorisation of the sample at all [101, 104], whereas other implementations facilitate movement of the sample in both lateral directions by a single coarse motion stage [105].

The electrical schematic and basic working principle of the STM is sketched on the left side of Fig. 2.2. In the STM presented in this work, the bias voltage is applied to the sample, and the tip is connected to a transimpedance amplifier which measures the tunnelling current. In the constant current mode, a feedback loop compares the measured current to a user defined setpoint and adjusts the vertical tip position accordingly to maintain the current setpoint. When scanning the tip over the sample, this feedback will cause the vertical tip position to follow features on the surface in order to keep the tunnelling current constant, so that a recorded map of vertical tip displacements resembles the topography — or, more strictly, the surface of constant tunnelling probability. Alternatively, the height of the tip above the sample can be fixed and the tunnelling current is measured as a function of the tip position. This is referred to as constant height mode, and is only useful for investigating very small surface variations as the extreme distance dependence of the tunnelling current in combination with the limited range and finite noise floor of the current amplifier limit the range of observable variations.

For STS studies, one needs to acquire spectra of the differential conductance with low noise. While it is possible to just record the current-voltage characteristics and calculate the first derivative, a quicker and more noise-resistant procedure involves the lock-in technique. For this, the bias voltage is modulated by a small alternating voltage $V = V_0 + V_m \sin(\omega t)$, so that the second order Taylor expansion of the measured tunnelling current about $V = V_0$ becomes

$$I\big(V_0 + V_m \sin(\omega t)\big) = I(V_0) + V_m \sin(\omega t) \cdot \left.\frac{\mathrm{d}I}{\mathrm{d}V}\right|_{V_0} + \frac{V_m^2 \sin^2(\omega t)}{2} \cdot \left.\frac{\mathrm{d}^2 I}{\mathrm{d}V^2}\right|_{V_0} . \tag{2.6}$$

The lock-in amplifier then multiplies the measured signal with the original modulation and integrates this product over time. For sufficiently large times T, the output signal of the lock-in then becomes

$$V_{\text{lock-in}} \propto \frac{1}{T} \int_0^T I\big(V_0 + V_m \sin(\omega t)\big) \cdot \sin(\omega t)\, \mathrm{d}t \tag{2.7}$$

$$\approx \frac{V_m}{2} \left.\frac{\mathrm{d}I}{\mathrm{d}V}\right|_{V_0} . \tag{2.8}$$

Using the lock-in technique to determine the differential conductance has two essential advantages over the numeric differentiation of the current-voltage dependence. First, only a narrow frequency band contributes to the output of the lock-in amplifier as Eq. 2.7 vanishes for all Fourier components of $I(t)$ with frequencies that do not match the modulation frequency ω. In STM, noise is usually not uniformly distributed across the frequency spectrum (white noise) but rather exhibits a $1/f$ behaviour. Therefore, it is preferable to shift the sensitivity of the measurement to a higher frequency, and filter all other undesired frequencies from the signal. Second, a topographic scan can be performed at the same time as a map of the differential conductance at fixed bias voltage is obtained if the modulation frequency is faster than the feedback loop. This mode of measurement allows to visualise the spatial variation of the LDOS at the surface at a specific energy.

2.2 Scanning tunnelling microscope induced luminescence

On a clean metal surface, the light emission is caused by the radiative decay of coupled, collective charge carrier oscillations in the tip and the sample, which are excited by inelastically tunnelling electrons. The emitted light is a measure of the inelastic fraction of the tunnelling current and the electromagnetic resonance properties of the cavity formed by the tip and the sample. For this reason, the interpretation of the acquired light signal requires separating two fundamentally different effects. First, the electromagnetic eigenmodes of the coupled electro-photonic system depend on the dielectric constants of the involved materials and the geometry of the junction, and are well described by classical electrodynamics. Particularly the tip shape has a large influence on the light emission but is difficult to assess or control, making it difficult to interpret individual spectra. Second, different materials on the sample provide different electronic states involved in the excitation process and thus affect the inelastic tunnelling probability. Finally, the tunnelling current and the cavity geometry are not independent of each other, as the tip distance is directly linked to the tunnelling current. This tip distance might change when an experiment is conducted at constant current due to lateral variations of the LDOS, and when the current is not held constant by the feedback loop, any variations of the tip distance dramatically affect the tunnelling current and, therefore, also the excitation rate of the gap plasmons.

2.2.1 Light emission from gap plasmons

Cavity resonance

In order to describe the optical response of the conduction electrons in the tip and the sample, it is satisfactory to consider a semi-classical approach in which the interaction of matter with the electric field is described by Maxwell's equations. In this framework, the response of the solid's electrons to the electric field — which inherently bears the signature of many different quantum mechanical effects such as the band structure and different scattering processes and transitions — is condensed into the complex, frequency-dependent dielectric function $\epsilon(\omega)$. The real part of this function describes

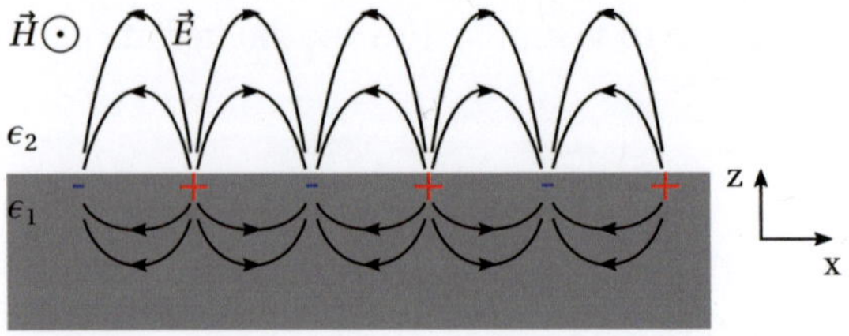

Figure 2.3: Electric field lines and surface charge distribution of a surface plasmon polariton (SPP) at a metal-dielectric interface. Adapted from [107].

the polarisability of the material, whereas the imaginary part describes energy dissipation. The optical properties of metals are largely determined by the conduction electrons, which can move freely except for the restoring force created when the local electron density does not compensate the positive charge of the atom cores. Even in a free electron gas, minimising the total electrostatic energy requires a homogeneous distribution of electrons, and any deviation from that results in a similar restoring force. Solving the classical laws of motion for the quasi-free electrons taking into account this interplay of inertia and restoring force results in oscillation eigenmodes, so-called plasma oscillations. For a free electron gas, the plasma frequency is [106]

$$\omega_P = \sqrt{\frac{ne^2}{\epsilon_0 m}}, \tag{2.9}$$

with the charge carrier density n and the electron mass m. In a metal, m actually denotes the effective mass of the electron which is determined by the band structure. When treated quantum mechanically, these charge carrier oscillations can be described as quasi-particles with a discrete excitation spectrum, which are referred to as plasmons.

The plasma frequency of a metal is a useful indicator to assess the optical response of a metal to an incident electromagnetic wave using the analogy of a driven oscillator. If the frequency of the external electric field is lower than the plasma frequency, the electrons will move 180° out of phase with regard to it, such that the real part of the dielectric function is negative. As a result, the electric field cannot penetrate the metal and is, instead, reflected. At the plasma frequency, resonance occurs, and if the frequency of the external oscillation is above ω_P, the electrons can finally no longer keep up due to their finite inertia and the material becomes transparent to radiation.

At the interface between a metal and a dielectric — or, in particular, the vacuum — the electric field generated by oscillating charge carriers extends into the dielectric. Considering the half-spaces $j = 1,2$ being occupied by the metal and the dielectric, respectively characterised by ϵ_j, and the interface being at $z = 0$, then a solution to Maxwell's equations without external driving force has the form [108]

$$\vec{E}_j = E_0 \begin{pmatrix} 1 \\ 0 \\ -k_x/k_{j,z} \end{pmatrix} \exp\Big(\mathrm{i}\big(k_{j,z}|z| + k_x x - \omega t\big)\Big), \tag{2.10}$$

with the wavevector components

$$k_x^2 = \frac{\epsilon_1 \epsilon_2}{\epsilon_1 + \epsilon_2} \frac{\omega^2}{c^2} \quad \text{and} \quad k_{j,z}^2 = \frac{\epsilon_j^2}{\epsilon_1 + \epsilon_2} \frac{\omega^2}{c^2}, \tag{2.11}$$

if $\epsilon_{1,2}$ are real. At optical frequencies, usually the conditions

$$\epsilon_1(\omega) \cdot \epsilon_2(\omega) < 0 \quad \text{and} \quad \epsilon_1(\omega) + \epsilon_2(\omega) < 0 \tag{2.12}$$

are fulfilled because the dielectric function of the metal ϵ_1 is negative and large compared to ϵ_2. As a result, k_x is real and k_z is imaginary, such that Eq. 2.10 describes waves with a maximum amplitude of the electric field at the interface, which propagate along the interface but exponentially decay in both half-spaces along the surface normal. The accordingly confined magnetic field oscillates in phase with the electric field and is parallel to the interface plane (transverse magnetic). Such a coupled mode of charge carrier oscillations in the metal and the electric field in the dielectric or vacuum is called a surface plasmon polariton (SPP) and sketched in Fig. 2.3. In case of a metal-vacuum interface, its energy is a factor of $\sqrt{2}$ lower than the energy of a bulk plasmon. Because of the — so far neglected — imaginary part of ϵ_1, k_x is also not entirely real, so that SPPs have a finite propagation length, which is in the order of several µm to several 10 µm depending on the wavelength and material [108]. Due to a mismatch of the dispersion relations and the conservation of parallel momentum, SPPs cannot directly couple to free-space photons on a flat surface. This is alleviated in the presence of surface roughness or — particularly in STM — a tip, which break the lateral translational invariance and, therefore, quasi-momentum conservation. Apart from that, also a prism can be used to mediate the coupling of propagating SPPs to the far-field [109].

When approaching two metals, the electric fields of their respective SPP modes overlap, and a new mode emerges in which the charge carriers of both metals couple with each other via the electric field in the gap. This new mode is characterised by opposite charges on the opposing surfaces as sketched in Fig. 2.4 and is called a gap plasmon, a term more commonly used in applications involving flat, sandwiched metal-dielectric-metal electrodes [110]. However, this also precisely describes the situation in STM, where the tip and the sample are in close proximity and separated by a vacuum

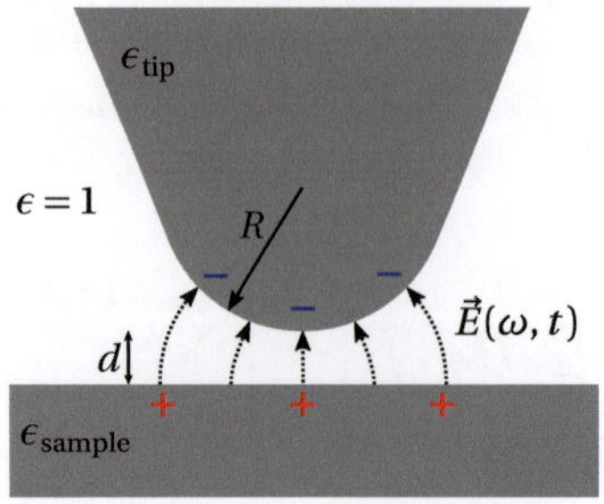

Figure 2.4: Lowest mode of a gap plasmon in the STM junction. Adapted from [46].

gap. The distinct difference is that in STM, the cavity consists of one flat electrode, i.e., the sample, and the apex of a sharp tip. This leads to a strongly enhanced field which is confined to a small volume around the tip apex. In a simplified geometry, i.e., a sphere in front of a plane metal surface, Rendell and Scalapino [111] assessed the confinement area of the electric field and found that it scales according to $\sqrt{2Rd}$ with the radius R of the sphere and the distance d from the surface. With typical tip radii well below 100 nm and tip distances below 1 nm, the optically active area is confined to an area far below the wavelength of a corresponding free-space photon, which is ultimately the key to the high lateral resolution of STML.

The gap cavity features distinct resonances ω_l if half a wavelength of the coupled mode equals the tip-sample separation [46]. This leads to a resonance effect at tip distances between 0.5 and 1 nm, with resonance frequencies ω_l described by the implicit relation [51]

$$\frac{1}{\epsilon(\omega_l)} = -\left(l + \frac{1}{2}\right)\sqrt{\frac{d}{2R}}, \qquad l = 0,1,... \tag{2.13}$$

if the tip and the sample are made from the same material. Evaluating this in the framework of the Drude model with the plasmon frequency ω_P and in the case $d \ll R$ this results in

$$\omega_l \approx \omega_P \sqrt{\left(l + \frac{1}{2}\right)\sqrt{\frac{d}{2R}}}. \tag{2.14}$$

From this, it is apparent that the shape of the tip has a dominant influence on the resonances, which was experimentally confirmed by the fact that spectra recorded on the same substrate with different tips often are very different, an issue which will be demonstrated in Sec. 4.2.1.

Another consequence apparent from Eq. 2.14 is that the resonance is red-shifted upon approaching the tip to the surface. This is qualitatively explained by Johansson [46]

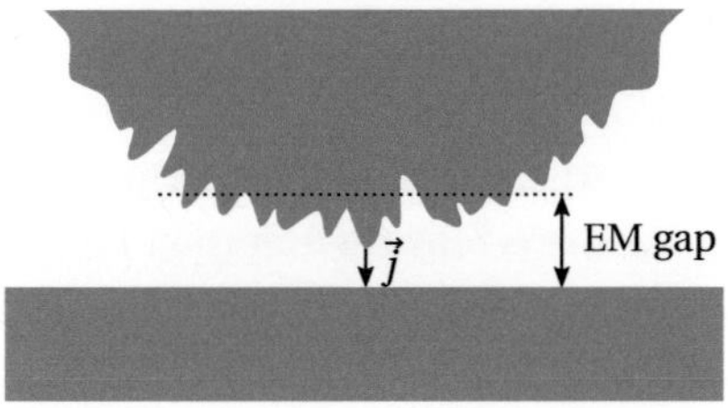

Figure 2.5: Multiple plasmonic resonances caused by a rugged tip. Adapted from [114].

by the opposite charges on the tip and sample in close proximity. The closer the electrodes get to each other, the more the opposite charges at the surface compensate each other, thereby reducing the restoring force acting on the charge carriers that have been displaced within the metals. In STM, however, only a small range of tip distances can typically be studied. If the tip is too close, it is subjected to a very high electric field which quickly damages the tip, and if it is too far away, the tunnelling current is so low that the induced photon emission is too weak to be detected. Within this accessible distance range, which spans a few 0.1 nm, the red-shift predicted by Eq. 2.14 was indeed observed by Aizpurua *et al.* [54]. The limitations of the classical approach leading to Eq. 2.14 are demonstrated by a few studies in which the authors managed to reduce the tip distance up to an atomic contact. Instead of the predicted infinite red-shift at zero distance, a moderate peak shift of 7 nm per 0.1 nm of tip displacement was observed [112, 113], indicating a coupling regime which is not well described by classical physics.

The lowest mode in Eq. 2.14 with $l = 0$ corresponds to a dipole oscillation and is the most prominent in STML. Higher orders are often not observed because the energy of the resonance approaches the energy of the SPP modes, such that the energy of the gap plasmon is carried away by a propagating SPP [46]. The lowest mode of a blunt tip, however, can be at low energy down to the infrared regime such that multiple peaks corresponding to higher orders are actually observable [115]. However, electroluminescence spectra in STML often feature multiple resonances, which cannot be explained in the simple model of a spheric tip. These resonances rather emerge from the non-ideal shape of real tips, which is often not accessible because the relevant part of the tip spans some 10 nm around the apex, and is difficult to characterise [116]. So far, two effects have been discussed in the literature which may result in these additional resonances [114]. First, the electric field is much less confined than the tunnelling current. This means, that a rugged tip as sketched in Fig. 2.5 will perform well for STM, because the tunnelling current only flows through the topmost extrusion. Within the reach of the electric field, however, there are several other so-called micro-tips which may host additional plasmonic resonances. It is noteworthy that blunt tips are particularly susceptible to this effect because of the larger interaction area. Another interpretation is that even a single micro-tip can host more than one resonance if it is asymmetric [114], each of which is associated with a different confinement length. In summary, blunt tips feature a denser spectrum which extends further to lower wavelengths compared to sharp tips, because they can feature higher-order modes and additional resonances caused by the tip roughness and asymmetry.

In an STML experiment, tuning the gap cavity is possible in several ways. The tip distance, for example, can be freely varied in STM. However, this is not independent from the other tunnelling parameters. Either the tunnelling current [30, 54] or the bias voltage [117] need to be adapted in order to vary the tip distance, and when interpreting the resulting variation of the light signal one needs to consider that both of these parameters also influence the emission. In Sec. 4.2.2, such an experiment involving a voltage variation at constant current will be presented and discussed. Apart from this, it was already discussed in the introduction how the topography of the sample can further modify the gap cavity.

Excitation

Gap plasmons can either be excited by incoming light or electrons. In STML, the latter is the case whereas different regimes have to be distinguished depending on the distance between the tip and the sample [32]. In the case of distances larger than 1 nm, a high voltage is required in order to achieve a significant current. This is the field-emission regime, which offers low spatial resolution, and the electron bombardment of the surface results in cathodoluminescence. At short tip distances below 0.2 nm, on the other hand, a very high tunnelling current already flows at low voltages. In this regime, luminescence is dominated by multi-electron processes due to the high rate of individual electron hopping events. Also, single-electron excitation is usually not possible in this regime as the energy of each tunnelling electron is not sufficient to excite plasmons if the tunnelling bias is low, which in turn is necessary in order to not damage the tip. The experiments in this work were performed in the tunnelling regime, i.e., at tip distances between 0.5 and 1 nm. This regime is characterised by intermediate voltages between 2 and 4 V and a tunnelling current of a few nanoampere. Luminescence at these parameters is caused by inelastic single-electron excitation of gap plasmons, whereas multi-electron processes are strongly suppressed because the lifetime of electrons in their respective final state is significantly lower than the average time-difference between individual tunnelling events.

In principle, two different excitation mechanisms are imaginable in this regime, either inelastic tunnelling or elastic tunnelling and subsequent recombination within the sample (hot luminescence). In STML, it was found that luminescence is exclusively caused by inelastic tunnelling [45]. The coupling of the tunnelling current to the charge carrier oscillations is mediated by shot noise. This mechanism was predicted early [111, 118] but experimentally proven only in 2010 [113]. Even though the tunnelling current is a direct current, the statistical distribution of time-differences between individual hopping events results in a broad, featureless power spectrum in the frequency domain. At zero temperature it is [111, 118]

$$\begin{aligned} \left|I(\omega)\right|^2 &= \frac{1}{2\pi R_0}(eV - \hbar\omega), \qquad && 0 < \hbar\omega < eV \\ &= 0 && \hbar\omega > eV, \end{aligned} \tag{2.15}$$

with the resistivity of the tunnelling junction R_0. This cut-off implies — in agreement with energy conservation considerations — that the energy of the tunnelling electrons must be higher than the energy of a plasmon mode in order to excite it. At finite temperatures, this cut-off condition smears out over $k_B T$, which results in the emission of photons with $\hbar\omega > eV$ [119], which is otherwise only observed in the regime of high tunnelling current as a result of correlated tunnelling of two electrons [112, 120]. Furthermore, higher bias voltages result in higher current fluctuation at the plasmon resonance, and thus in a higher excitation probability. At constant current, however, higher bias voltage is accompanied by increased tip-distance, which counteracts the increased excitation probability due to reduced coupling of the charge carriers across the gap.

The tunnelling current essentially consists of two different channels which are its elastic and an inelastic portion. Both, the elastic and inelastic tunnelling probability scale with the LDOS of their corresponding final states. This means that if the LDOS of the final states of the inelastic channel varies then the inelastic tunnelling probability varies as well. These variations are negligible in a traditional STM experiment because the vast majority of the tunnelling current is elastic. In STML, however, this variation can directly be probed because different plasmon frequencies will consequently be excited with a different probability. This is particularly notable if the surface features spatial variation of the LDOS due to adsorbants or different materials because the light emission will vary with the tip position [29, 121, 122]. This is the main contribution to the variations of the light emission presented in Ch. 5.

Emission

The lifetime of localised plasmons is limited by resistive losses in the materials, i.e., due to the imaginary part of $\epsilon(\omega)$, to a few 10 fs [123], whereas the average lifetime of radiative transitions is in the order of 200 fs [70]. For once, this provides the opportunity to use the emitted light as a fast probe for the tunnelling current [70, 71]. Most importantly, however, the high dissipation-related decay rate significantly broadens the resonance and causes most of the gap plasmons to be dissipated. Apart from this direct quenching, another significant fraction of gap plasmons decays into propagating SPPs [124–126], which in turn cannot contribute to the light emission anymore unless they are deliberately coupled to a prism or a grating. As a result, about 1 % of the tunnelling electrons excite a gap plasmon [46], but only a small fraction of those plasmons emits a photon.

Still, the electron-to-photon conversion in STML is orders of magnitude higher compared to the case of flat electrodes [19] due to the tip acting as an antenna for coupling of the near-field to the far-field, an effect which is similarly utilised to achieve sub-wavelength resolution in tip-enhanced near-field optical microscopy [127]. Apart from that, the substantial electric field at the tip apex provides strong coupling to the tunnelling electrons [128]. As the electric field enhancement is directly related to the tip-sample distance, changes of the latter in the order of less than 0.1 nm cause significant variations of the amplitude of the electric field at the surfaces [46, 129]. As a result, the electron-to-photon conversion efficiency notably increases when approaching the tip to the surface even by tiny steps [30].

Similarly, the tip radius not only affects the resonance frequency but also the brightness of the emitted light because sharp tips result in a higher field enhancement. In a theoretical work of Aizpurua *et al.* [130], a more realistic, hyperbolic tip model with separate opening angle and tip radius was considered, and the authors found that the aperture of the tip was mostly responsible for the overall shape of the spectrum, whereas the apex radius dictates the intensity and sharper tips produce a brighter signal at constant current. Tips with a radius below 30 nm, however, are expected to yield less intensity again, as the tip area must not be too small in order to function as an antenna and effectively mediate the emission [46]. Therefore, a tip radius of 30 nm is expected to be optimal with regard to the emission intensity. In practise, however, it is impossible to experimentally control the radius of the tip apex so precisely. But an essential conclusion of all these considerations is that the tip shape largely dictates the brightness, number and energy of resonances, and material properties often play a secondary role. Therefore, all STML experiments need to be devised in a comparative fashion, i.e., the *variation* of the emission spectrum depending on surface features or tunnelling conditions potentially allows conclusions about underlying effects, whereas a single spectrum has little informative value.

The tip and sample materials, after all, also affect both the resonance and emission intensity. In order to study plasmon-mediated light emission, usually silver, gold and copper are used as samples due to their low damping. Silver, in particular, also features an almost ideal free-electron like behaviour in the optical frequency range due to the high onset of interband transitions [26]. Gold or silver are also preferred as tip material, as opposed to tungsten, which is much more popular for regular STM studies. When compared to tungsten, the achieved field enhancement with gold or silver is higher and the resistive damping is lower [46, 129, 131], which results, on average, in one order of magnitude brighter emission spectra [30].

Theoretical estimates and experimental reports on the photon generation efficiency on noble-metal surfaces are consistently between 10^{-4} and 10^{-3} [20, 25, 26, 45–47], depending on the tip and sample materials and, of course, the tip geometry. In one case, even an exceptionally high value up to 10^{-2} has been reported [132]. On semiconductor surfaces, estimates for light emission from minority carrier injection are generally one order of magnitude lower [34], and can be as low as 10^{-7} [133].

For the lowest oscillation mode, Rendell and Scalapino [111] calculated the angular emission distribution and found a radially symmetric emission maximum which is tilted away from the surface at an angle of 35°. This is particularly interesting for optimising the collection optics.

2.2.2 Light emission from molecules

Even though molecular studies are beyond the scope of this work, the STM designed and presented in Ch. 3 was specifically build for that purpose. For this reason, a short summary will be presented how the presence of a molecule in the STM junction affects the light emission.

In general, radiative decays have longer lifetimes than channels related to electron-electron scattering. If, in the simplest case, a molecule does not have long-lived excited states, or they cannot be excited by the tunnelling current due to inappropriate tunnelling conditions, then it still modifies the LDOS and the current-distance characteristics. At constant current, therefore, the tip-distance is often increased which results in a reduced gap plasmon coupling and lower intensities. Early reports of molecular resolution in STML were ascribed to this effect, and the molecules merely acted as a dielectric spacer and tuned the gap plasmon [134, 135]. Similarly, molecules affect the voltage-current characteristics. The electric field in the gap shifts molecular orbitals relative to the Fermi edges of the tip and the sample. When these orbitals are appropriately aligned relative to the Fermi edges of the electrodes, resonant tunnelling occurs and the tunnelling current rises significantly. Accordingly, this excites more gap plasmons, which means that the molecular electronic structure is utilised as a voltage-controlled switch for the plasmonic light emission [136].

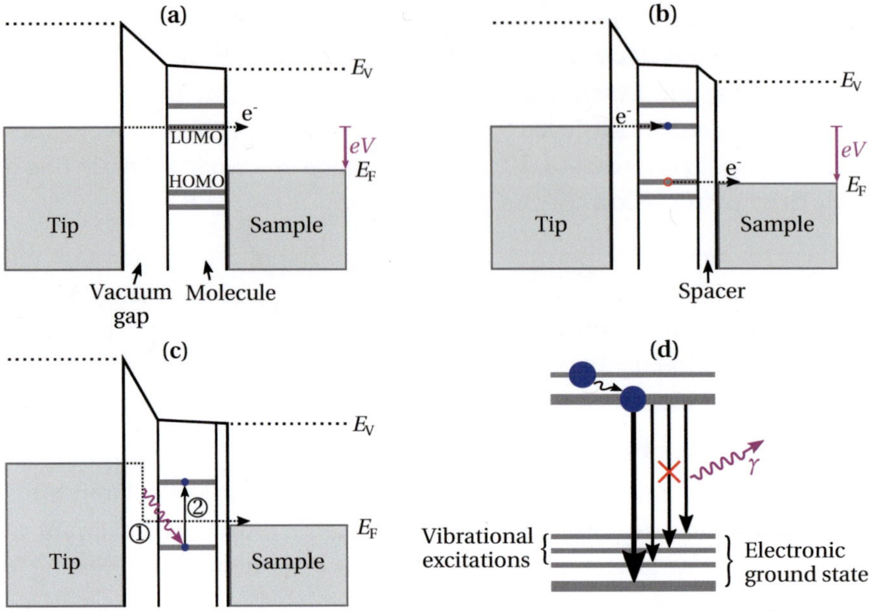

Figure 2.6: Excitation and de-excitation of molecules inside the STM junction. (a) and (b) show basic energy diagrams with and without decoupling from the substrate. Electrically excited molecular states can only be prepared in (b), as indicated by an electron (blue dot) in the LUMO and a hole (red circle) in the HOMO, which may later recombine. Adapted from [75]. (c) Indirect excitation of electrons from the HOMO into the LUMO by gap plasmons, adapted from [77]. (d) Possible de-excitation channels when the molecular energy landscape features different vibrationally excited end states.

In order to actually achieve fluorescence, one needs to excite a molecule to a long-lived excited state, which first and foremost requires decoupling from the metallic substrate. Otherwise, hybridisation occurs and the electron in the excited state is efficiently transferred to the substrate. Decoupling also facilitates proper alignment of the molecular orbitals between the Fermi edges of the electrodes, which is required to excite a molecule in the first place [75, 137]. This mechanism is illustrated in Figs. 2.6(a) and (b), depicting the molecule between the metallic electrodes. The molecule is characterised by few discrete molecular orbitals, and the two most relevant orbitals are referred to as highest occupied molecular orbital (HOMO) and lowest unoccupied molecular orbital (LUMO). Without a spacer (Fig. 2.6(a)), the molecular orbitals are pinned to the substrate, and the voltage drop across the molecule is minimal. If the bias voltage is increased, the molecular orbitals can be aligned to the Fermi edge of the tip such that resonant tunnelling occurs, but this does not excite the molecule because the highest occupied molecular orbital is always below the Fermi edge of the sample and, therefore, always fully occupied. In Fig. 2.6(b), the molecule is decoupled from the sample and the voltage drop across the gap is distributed among the vacuum gap between tip and molecule, and the spacer between molecule and substrate. Therefore, the orbitals of the molecule can shift with respect to *both* electrodes, and by applying an appropriate bias voltage, a situation can be achieved in which the HOMO is slightly above the Fermi edge of the sample, and the LUMO is at or slightly below the Fermi edge of the tip. Therefore, holes will form in the HOMO, and the LUMO will be occupied with electrons from the tip. This is an electronically excited state which has the potential to decay by emitting a photon.

More recent studies have found another indirect excitation mechanism, shown in Fig. 2.6(c). Here, the tunnelling electron first excites a gap plasmon, which in turn excites an electron in the HOMO [77, 83, 84, 138]. That way, fluorescence is also achieved when the decoupling is not sufficient to lift the HOMO above the Fermi edge. It is still required to protect the molecule from hybridisation.

However, molecules do not only have electrically excited states, but also feature vibrational degrees of freedom which can be excited, which adds to the observable transitions. The energy of vibrational excitations is typically much lower than the energies of electrical excitations, so that they can be grouped as shown in Fig. 2.6(d). The different sets of lines belong to the same electrical states, whereas only the respectively lowest line is also the vibrational ground state. The excited molecule may at first be in an electrically *and* vibrationally excited state, but, if so, quickly relaxes to the vibrational ground state [79, 137]. From there, then several end states are available for the charge recombination, whereas some of these transitions are suppressed according to the Franck-Condon-Principle, which is illustrated by the crossed transition. Because each of these transitions has a different energy, the emitted photon wavelengths bear the vibrational fingerprint of the molecule [137, 139], which has been demonstrated even with spatial resolution in the light emission [77].

In the literature, a wide variety of decoupling techniques has been employed so far. This includes spacer layers consisting of C_{60} molecules [136], oxides [79, 137], or NaCl [75, 77, 81, 82, 84], or the molecules were deposited on a wide-gap semiconductor [140, 141]. Fluorescence was also demonstrated with tailored molecules that consist of a chromophore integrated within a molecular wire, which can be suspended between the sample and the tip, thereby achieving decoupling of the optically active centre by the leads of the molecule itself [83, 142].

It is noteworthy that the light emission from single molecules is strongly enhanced by the presence of the STM tip, which is one of the reasons why STML studies of single molecules are so appealing. The gap cavity increases the light emission probability of excited molecules because the local photonic density of states ρ_μ is drastically increased due to the plasmon modes [73]. By applying a dipole approximation and using Fermi's Golden Rule, the spontaneous decay rate γ of any two-level quantum system is [108]

$$\gamma = \frac{\pi\omega}{3\hbar\epsilon_0}|\vec{\mu}|^2\rho_\mu(\vec{r},\omega), \tag{2.16}$$

with the transition dipole moment $\vec{\mu}$. This effect is referred to as tip-enhanced fluorescence and is similarly utilised in tip-enhanced Raman spectroscopy [143].

2.2.3 Experimental implementation

While it is *possible* to just put a detector close to the STM junction in order to collect the emitted light [20, 146, 147], sufficient detection rates for significant spectra — preferably with spatial resolution — can only be achieved by covering a high solid angle around the tip. The difficulty of this depends on the experimental conditions. A few STML setups operate in air [56, 146–149], where the collection optics is easy to install and focus, and many others operate in UHV at room temperature. This already complicates the matter because tips and samples need to be changed *in-situ* and the accessibility of the junction is limited in UHV. Single molecule studies, however, also require low temperatures. So far, only the ten STML setups listed in Tab. 2.1 have been reported which operate at or below 10 K and in UHV, and none of these operates below 4 K. It is apparent that the achieved detection rates do not necessarily scale with the covered solid angle because different detection techniques were used, the emission intensity depends on the tip, and due to unaccounted for losses along the light path between the collection optics and the detector.

Table 2.1: Low-temperature STML setups operating in UHV reported in the literature so far. Empty fields indicate missing information in the respective publications.

Year	Ref.	Collection	T (K)	Coverage (%)	Max. counts per nC
1995	[144]	Transparent tip	10	n.a.	10000 (GaAs)
1999	[87]	Parabolic mirror	4	50	20 (GaAs)
2002	[47]	Lens	5	9	50000 (Ag)
2003	[137]	Lens	8		
2009	[145]	Lenses (2)	5	4.8	3000 (Au)
2010	[86]	Lenses (3)	4.5	27.6	
2014	[76]	Parabolic mirror	5		
2016	[83]	Lens	4.5	2.5	6600 (Au)
2016	[82]	Lenses (2)	8	20	20000 (Ag)
2016	[81]	Lenses (2)	4.6	16	

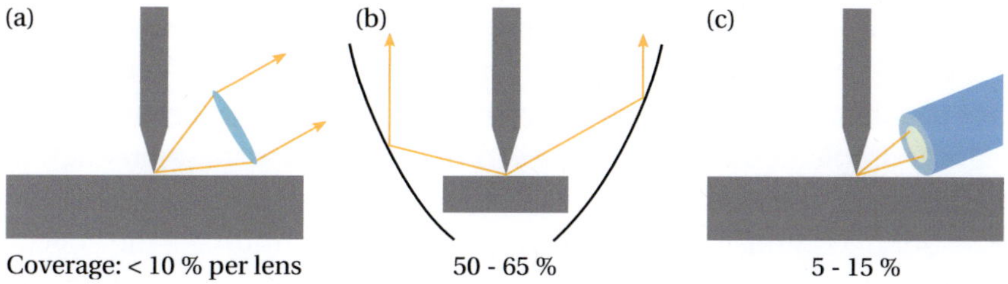

Figure 2.7: Popular implementations of the primary collection optics using (a) one or several lenses, (b) a parabolic or ellipsoid mirror or (c) one or several optical fibres or fibre bunches close to the junction. Adapted from [31].

Most designs implemented a collection optics close to the junction, which collimates the light and routes the collected light to the detector where it needs to be focussed again. If the STM is mounted inside a UHV cryostat, then the collected light also needs to be routed through the cryogenic shields and a UHV viewport, thereby complicating the light-path even further. The most popular primary collection optics is one or several lenses due to the relative simplicity of the design [22, 23, 47, 75, 81, 82, 86, 132, 137, 145, 150, 151]. Since recently, there is even a commercial STML setup available from *Unisoku*, which also uses a lens for collecting light. Also, curved mirrors have been employed in two variants, namely parabolic [67, 76, 87, 152] and ellipsoid [25, 153] mirrors. Parabolic mirrors collimate the light similarly to lenses and require refocussing, whereas ellipsoid mirrors focus the source directly onto another focal point. Curved mirrors easily cover a very large solid angle, however, they need to be precisely adjusted.

The key disadvantage of lenses and mirrors is the free-space routing between the collection optics and the detector, which requires the collection optics to almost perfectly collimate the collected light. This requires to mount the lenses or mirrors on micrometre screws or piezo-based positioning systems with many degrees of freedom in order to align the focus with the tip apex, at least after every tip change. As soon as for example several lenses are involved, the complexity and size of the required contraptions quickly escalates. Only few designs so far attempted a fixed collection lens [83, 145], thereby accepting the inevitable collection losses. Particularly troubling is also the limited long-term stability of optical pathways where the components are far apart and at different temperatures, i.e., some are installed inside and others outside of a cryostat. Then, the tip tends to drift out of the focus over the course of hours, rendering long-term measurements as presented in Ch. 5 of this thesis essentially impossible [47].

An alternative to lenses and mirrors are optical fibres close to the junction [148, 154–157]. An appealing aspect of this idea is that it requires no free-space routing and the fibre can be guided through the cryostat alongside the electrical cabling. However, an optical fibre needs to be either quite large or extremely close to the junction in order to cover a significant fraction of the solid angle. Then, they need to be retracted by a significant distance and later repositioned for each tip or sample change. And while it is possible to implement several fibres in order to increase the solid angle coverage, it is not efficient to mount all of them on the same positioning system [156], but it is rather favourable to position each fibre individually to optimise the collection efficiency. Even if the problem of positioning the fibres is finally solved, they transmit only light entering in a small cone. The STM tip, however, is a point source with a large angle distribution, which fundamentally limits the usability of fibres. Another slight disadvantage of many fibre-based designs is that the polarisation and angle distribution of the light is not preserved, whereas the polarisation of the emitted signal conveys information about the angular momentum of the involved excited state due to optical selection rules [158] and might potentially be of interest, even though only few studies actually make use of this information.

Apart from the popular solutions already discussed, also two more exotic designs have been realised so far. Murashita *et al.* [144] devised a conductive, transparent tip coupled to an optical fibre which yielded a respectably intense signal (see Tab. 2.1) but is also by design limited with regard to the collected signal. Another approach is to mount a large microscope objective below a transparent sample [64, 124, 149], which however heavily limits the available sample choices and is entirely incompatible with a cryogenic setup. Particularly interesting is also the approach of Khang *et al.* [87] to combine the STM tip and a parabolic mirror to an integral unit. The alignment of the mirror relative to the tip can be ensured during fabrication, and the combined assembly can be swapped as a whole, rendering *in-situ* focussing unnecessary. A similar approach will be followed in this work.

Apart from the efficiency of the primary collection optics, the light signal also suffers from additional losses in the detection setup. As a result, the total detection efficiency of many designs is in the order of a few percent. Achieved count rates that were disclosed in publications so far are usually spread between a few thousand and several ten thousand photons per nanoampere of tunnelling current and seconds of exposure. Tab. 2.1 gives a partial overview over the achieved detection rates in low-temperature setups published so far. Two extraordinarily efficient setups with detection rates of 1 to 4×10^6 photons per nC have also been reported [132, 152]. However, these numbers are difficult to compare. The quantum efficiency depends very much on the tip and is unknown, most experiments have been performed on different substrates, and many setups employ photon counters instead of spectrographs, which are much more efficient but lack any spectroscopic resolution. The only really comparable figure of merit specific to STML setups is the wavelength-dependent total collection and detection efficiency, and this is difficult to assess and rarely disclosed.

2.3 Direct laser writing

Photopolymers are a class of materials that undergo structural changes when interacting with light, in particular because of the formation of reactive radicals that cause chemical interlinking of the constituents. By that, a liquid photopolymer — commonly referred to as a *resist* — can solidify. Usually, the formation of radicals requires ultraviolet light due to the involved bond-energies, but in some resists the same reaction can be triggered upon correlated absorption of two infrared photons. The probability for this scales with the square of the intensity of the infrared light, and exhibits a strong threshold behaviour. In DLW, whose basic principle is sketched in Fig. 2.8, this is utilised to selectively solidify the resist only within a small ellipsoid around the focus of a laser beam, because only there, the square of the light intensity is sufficiently high to trigger the cross-linking reaction [88, 159, 160]. The size of this ellipsoid, which is also referred to as *voxel*, is only diffraction limited. For the structures presented in this thesis, it has a diameter of 200 nm and a vertical extent of 500 nm. Moving the laser focus through the resist creates a continuous line of solidified material, that is extended to two dimensions by placing lines next to each other such that they overlap. This can be extended into the third dimension in order to write almost arbitrary three-dimensional objects layer by layer, again such that the individual layers overlap. This inevitably results in a stepped surface instead of a perfectly smooth one. Subsequently, the liquid resist is developed, i.e., dissolved in a solvent in order to expose the solidified structure.

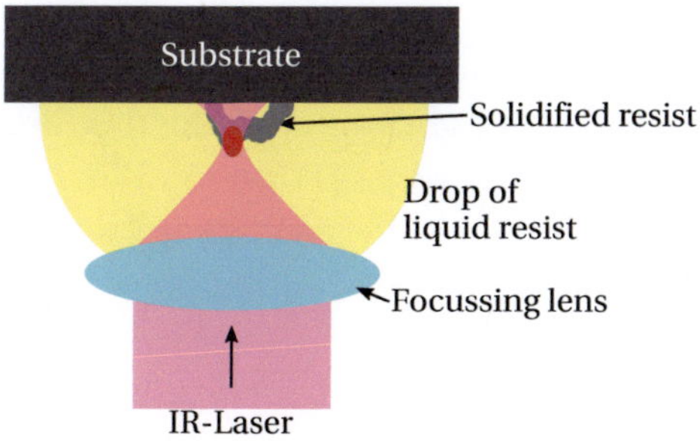

Figure 2.8: Direct laser writing in the so-called dip-in configuration, in which the focussing objective is immersed in the liquid resist. Cross-linking only occurs in the focus of the laser beam (red ellipsoid), which is swept through the resist to solidify the material along a pre-programmed trajectory.

This modern, additive fabrication technique offers almost unlimited freedom of design for creating small devices with very high resolution in the sub-micrometre range. It has, among others, been used to write photonic crystals and metamaterials with artificial optical and mechanical properties [161–163], plasmonic devices [164], as well as custom reflective and refractive micro-optics [165–168]. The technology has been commercialised by *Nanoscribe* in 2007 and has ever since attracted much attention due to its diverse application range throughout optics, medicine, fluidics, and micromechanics.

2.4 Surface states on metal surfaces

Electronic wavefunctions in a solid have the form of a Bloch wave with real momentum components, which freely propagate within the bulk and are finally reflected at the surface. At the surface, however, additional solutions to the Schrödinger equations emerge, which are not always matched by states in the bulk regarding symmetry and quantum numbers, and can therefore neither penetrate into the crystal [106, 169] nor escape to the vacuum because they are in a bound state inside the coulomb potential generated by their own image charge. The latter leads to a tail of the coulomb potential which decays according to $1/x$ into the vacuum. Accordingly, these states are referred to as image states or Shockley states [170]. For the complex wavevector component along the surface normal, this potential results a Rydberg-like series of discrete solutions, as sketched in Fig. 2.9. The wavevector components along the surface remain real and the carriers exhibit a free-electron behaviour characterised by a parabolic dispersion. For most metals, the image state with lowest discrete eigenenergy is partially filled and can be observed as a two dimensional electron gas which is confined to the surface [171, 172].

These surface states appear in STM images and particularly in maps of the differential conductance in the form of characteristic interference patterns due to scattering at surface impurities and step edges [173–175]. Wherever the wavefunction of the surface state must vanish, incoming waves are reflected and interfere with themselves, resulting in a standing wave pattern referred to as Friedel oscillations.

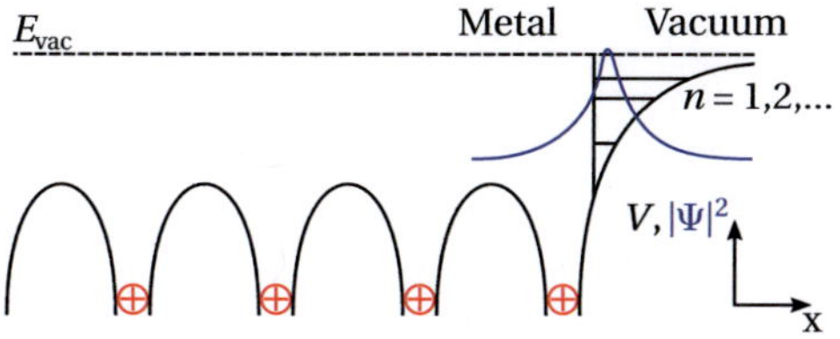

Figure 2.9: Coulomb potential at the metal-vacuum interface, giving rise to image states that are confined to the surface. Adapted from [106].

3 Experimental setup

Scanning tunnelling microscopes with light collection capability are still quite rare. To study electroluminescence from the tunnelling junction, an entirely new STM was designed and built, which was titled *γ-STM*. Almost every aspect of the experimental setup was designed using the computer-aided design (CAD) software *Solid Edge* by *Siemens* before the corresponding components were manufactured, purchased or assembled. This way, expensive and time-consuming mistakes were largely avoided. Additionally, this software features a module for simple finite element simulation, which was used to optimise the rigidity of the STM design. Most custom parts were then fabricated by the mechanical workshop of the *Physikalisches Institut*, and others by the mechanical workshop of the *Institute of Nanotechnology*, both at the KIT. Some custom electronics were provided by the electronic workshops of both aforementioned institutes, others were home-made.

In the scope of this work, existing plans of the UHV chamber were modified, and the STM including the light collection as well as the radiation shields with the so-called parking mechanism were designed from scratch. Then, a former storage room at the *Institute of Nanotechnology* was converted into a laboratory and the entire UHV chamber including the cryostat and the STM were assembled. During the commissioning phase, the system was optimised and thoroughly characterised. All important aspects of the design as well as the performance tests presented in the next chapter have been published in the journal *Review of Scientific Instruments* [176]. This work was supported by Moritz Winkler [177] and Lars Wilmes [178] in the scope of their master's theses. Moritz contributed to the design of the light collection and helped with the assembly of the STM, and Lars assisted me during the commissioning phase and during first successful experiments.

In the following chapter, the experiment will be presented from the inside out: starting at the mirror tip and the light collection setup, to the custom STM, the cryostat and finally the enclosing UHV chamber.

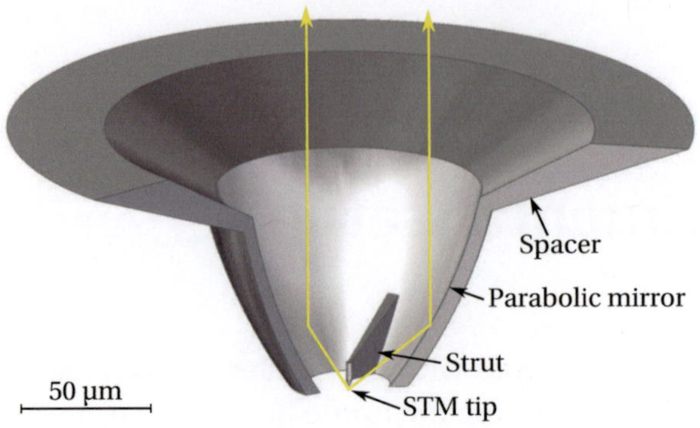

Figure 3.1: Sectional view of the mirror tip design and sketched light trajectories originating from the tip.

3.1 Light collection and detection

The STM tip can be approximated as a point source that emits light into a large solid angle. Optical components for collecting the light therefore should cover as much of the solid angle around the tip as possible. In this work, a design was chosen in which the STM tip is suspended at the focus of a nearly fully enclosing, microscopic parabolic mirror which collimates the light from the STM junction into an optical fibre located directly above the mirror. The fibre is fed through the cryostat, outside the UHV and into a grating spectrometer. Thus, as many photons as possible are collected from the tunnelling junction and guided to the detector.

3.1.1 Mirror tip

The core innovation of this design is to integrate the STM tip into a light collecting parabolic mirror fabricated as one piece by DLW that will be referred to as *mirror tip* and is presented in Fig. 3.1. The design, fabrication and characterisation of these mirror tips will be described in the following.

Design

To suspend the STM tip inside a parabolic mirror, it needs to be fixed in such a way that the fixture obstructs as little of the light path as possible. At the same time, this fixture needs to be rigid to not compromise STM operation. Hence, it is best to hold the tip by directly connecting it to the inside of the mirror with a short strut, as depicted in Fig. 3.1. It is not feasible to design this as a multi-part assembly, because any fasteners and adjustment components would add to the size of the fixture inside the mirror and obstruct the light path. Therefore, it is favourable to manufacture the arrangement of mirror, strut and tip in one piece.

Conventional fabrication techniques are not capable of creating such a geometry, especially not at a scale small enough to collimate the light to the small diameter of an optical fibre. Using DLW, however, it is possible to overcome these limitations and fabricate almost arbitrary shapes at the micrometer and sub-micrometer scale with high precision and small surface roughness [160, 167]. The two essential components of the design, namely parabolic mirrors [165, 166, 179] and sharp tips [180] — although in that case used for atomic force microscopy instead of STM — have already been fabricated with this technique.

The design of the mirror tip as shown in Fig. 3.1 consists of four parts: the STM tip, the strut holding it, the parabolic mirror surface and a spacer. When the STM tip is in tunnelling contact, it emits light, which is reflected at the mirror surface and supposed to exit at the top of the mirror as a directed and parallel beam. This assumes geometric optics which is well justified considering that low output beam divergence has already been demonstrated with way smaller parabolic reflectors [165, 166]. At the bottom end, the opening of the mirror has a diameter of 30 µm, which is much larger than the wavelength of the emitted light. The height of the parabolic section is 80 µm, opening up to a diameter of 100 µm at its top.

When the mirror tip is in contact with a sample, the tunnelling current is supposed to flow through the tip and not through the bottom edge of the mirror. However, the STM tip should not protrude from the bottom of the mirror too far in order to maximise the solid angle coverage. In STM, only very flat samples are investigated, therefore the tip was chosen to protrude by only 1.4 µm. This results in a safety clearance angle of about 5°, which is expected to never be exceeded by the sample roughness and tilt between tip axis and sample plane normal. To keep the obstruction of the light path to a minimum, only one strut is holding the tip. It was verified that this is rigid enough for STM operation by performing a finite element simulation, that determined the resonance frequency of the suspended tip at 1 MHz. Considering the open rim at the bottom of the mirror and shadowing by the tip and the strut, the covered solid angle of the light collecting mirror amounts to 75 % of 2π, which is to my best knowledge unmatched by any other design reported so far, and even by a large margin if only low-temperature UHV setups are considered. Assuming the angular intensity distribution reported by Rendell and Scalapino [111], this figure even increases to 82 %.

The purpose of the spacer at the top of the design in Fig. 3.1 is to separate the delicate mirror shape from the substrate on which the mirror tip is written, which is necessary for several reasons. First, the DLW resist shrinks a little during the development process [181, 182]. At the top end, it is fixed to the substrate though, resulting in distortion close to the interface. The same is to be expected as soon as the mirror tip is deployed in the cryostat due to different thermal expansion coefficients of the polymer and the substrate. Second, the substrate is reflective as will be discussed in Sec. 3.1.1, rendering it impossible to write features close to the interface with high resolution. In the design, the spacer lifts the mirror shape by about 30 µm, raising the total height of the structure to 110 µm. It also opens up to an inner diameter of 185 µm because the mirror will be centered around a pinhole in the substrate that is between 130 and 150 µm wide and it is advisable to stay clear from the major and unavoidable substrate inhomogeneities at the edge of the pinhole during the writing process for reasons explained in the next section.

DLW substrate and writing process

It is convenient to write the DLW structure onto a substrate, to which it will be permanently attached. Otherwise, handling the finished structure would be very difficult, and the structure also needs to be fixed during the writing process. So even after fabrication, this substrate will remain an inseparable component of the tip assembly and will also serve to suspend the mirror tip in front of the optical fibre. Once installed in the STM, the opening of the mirror should be seated very close to the fibre end. Therefore, the DLW substrate — which inevitably separates the opening of the mirror from the fibre end — should be very thin, yet rigid, and feature a pinhole to clear the light path. It also needs to be electrically conductive to carry the tunnelling current. For these reasons, it was chosen to write the mirror tips on a 25 µm thick steel sheet, which is welded onto a tip holder discussed in Sec. 3.2.3. A pulsed laser with a focus size of 20 µm is used to cut the steel sheet to a disk with a diameter of 2.2 mm and a pinhole at its centre, see item 3 in Fig. 3.10(a). Once welded onto the tip holder, this thin disk is supported along a ring with an inner diameter of 1.6 mm. Owing to these small dimensions, the axial resonance of the assembly is above 100 kHz despite the low thickness of the steel sheet, which is by far rigid enough for STM operation. The laser cutting was performed by Dr. Thomas Reisinger at the *Institute of Nanotechnology* and proved more convenient than fabricating the steel sheet discs manually, which is also possible but labourious.

For DLW, metal is not an ideal substrate. It reflects the laser beam during the writing process, adding to the intensity at the laser focus. This focus intensity must be kept within a certain range for the polymerisation to initialise without burning the material. If a reflective surface cannot be avoided (chemical blackening of the surface was tried and proved impractical) it should at least be uniform across the writing area, such that a source intensity can be found for which the focus intensity stays within the required range throughout the writing process. This is troublesome, because the laser cutting leaves a small burr and the general roughness of any commercial steel sheet is considerable. Thus, the steel foil was polished to a mirror finish prior to the DLW process.

The DLW process was performed by Dr. Martin Schumann and Christian Kern in the group of Prof. Martin Wegener at the *Institute for Applied Physics* with a commercial *Nanoscribe* Photonic Professional GT laser lithography system and involves two steps with different resists. It is desirable to fabricate the intricate paraboloid and especially the tip, which is supposed to be very sharp in the end, with a resist that facilitates high resolution, small feature sizes and small surface roughness, such as *Nanoscribe* IP-Dip. However, the workable focus intensity range of this resist is so small that it is not favourable to use this resist close to the reflective substrate interface. Instead, the spacer of the mirror tip is written with a resist offering a higher intensity tolerance, namely *Nanoscribe* IP-S exposed with a 25× objective with a numerical aperture (NA) of 0.8. After the spacer has been written and developed, IP-Dip is applied and the rest of the structure is written with high resolution using a 63× objective with a NA of 1.4. Both cycles were written with the lens of the writing system being immersed in the resist.

As described in Sec. 2.3, the layer-wise fabrication process of DLW inevitably causes a stepped surface instead of a perfectly smooth one. The vertical distance of these individual layers was optimised to achieve a reasonably smooth surface without adding too much to the writing time. It was found to be 200 nm optimally. Also the volume of the mirror walls is not completely filled with DLW trajectories to reduce the writing time. It is rather made up of a semi-hollow scaffold, providing sufficient mechanical strength. This can also be seen at the bottom of the mirror in Fig. 3.2(c). The total writing time of both steps then amounts to approximately 30 minutes.

Coating

At this point, the mirror tip consists of a crosslinked polymer and has to be metallised in order to be reflective at the mirror surface and conductive all the way from the substrate to the tip. First of all, the parts are coated with 20 nm of Al_2O_3 using atomic-layer deposition. This acts as an adhesion layer for the further metallisation, which is then done from both sides consecutively using vapour-deposition *in vacuo*: the tip side is coated from below and the mirror side from above through the pinhole in the steel sheet substrate. This requires a rotatable manipulator with no obstruction below or above the tip, which is possible in the load-lock chamber (see Sec. 3.4). To achieve maximum reflectivity, the mirror side was coated with 100 nm of silver. Accounting for the steep angles of the geometry being coated, actually 200 nm to 300 nm of silver had to be deposited onto the mirror side in order to reach the desired thickness. The tip side is also covered with silver which is particularly suitable due to its high plasmonic response. Adhesion of the metal coating on the tip side is, however, critical due to the stress the coating is subjected to at the STM tip caused by tip-sample interactions and the high electric field at the tip apex, especially at the elevated bias voltages required for photon emission in the visible range. Therefore, 20 nm of chromium are deposited on the tip side as an additional adhesion layer before finally depositing 60 nm of silver. Mirror tips that have only been covered with silver or only chromium and silver have suffered from loss of conductivity during tunnelling rather quickly, which can be attributed to the metal coating being easily torn off.

Characterisation

After the metallisation, some of the produced mirror tips were characterised in a scanning electron microscope (SEM), partly in the group of Prof. Wegener and partly with the assistance of Simone Dehm at the *Institute of Nanotechnology*. Representative micrographs of different mirror tips are presented in Fig. 3.2. Of special interest are the achievable sharpness of the tip and the smoothness of the mirror surface, as these two properties determine the properties of the light emission and the collimation performance, respectively.

The apparent tip radius in Fig. 3.2(b) is 100 nm. For good STM performance a sharp tip is of course desirable, and that is even more the case for STML experiments, as tips with a smaller tip radius R are brighter compared to blunter tips at the same tunnelling current [130], and also provide a more well defined luminescence spectrum with fewer peaks. Sharper tips also confine the gap plasmons to a smaller area and, therefore, allow for a higher resolution when investigating lateral variations of the cavity properties. In STML experiments, often tip radii well below 30 nm have been reported so far, which were determined by comparing calculated with experimental emission spectra [26, 114–116]. This, at first, seems to oppose the usability of the tip presented in Fig. 3.2(b). However, only the topmost atoms of the tip determine the achievable STM resolution, and the portion of the tip that affects the light emission also extents no more than a few nanometres from the apex [116] and cannot be resolved in SEM. For once, this microscopic apex might be much sharper than the macroscopic radius as seen in SEM to begin with. But even if that is not the case, the tip can still be worked *in-situ* by applying voltage pulses and dipping it into the surface, until the desired luminescence properties are achieved. This is a standard procedure for sharpening STM tips *in-situ* [183] and is most often also necessary when using electrochemically etched metal tips for STML measurements [32, 184].

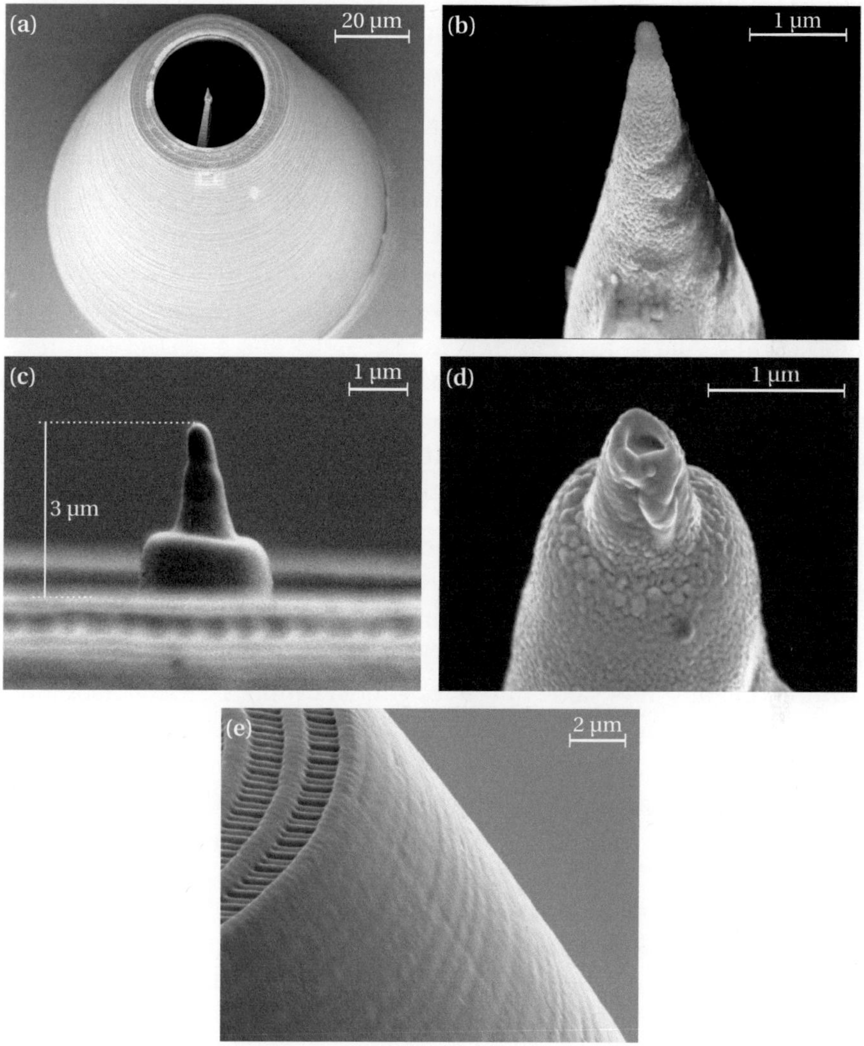

Figure 3.2: Fabricated mirror tips after metallization as inspected by an SEM. (a) Overview of the upside down structure with the STM tip pointing upwards, (b) close-up SEM image of the STM tip close to the focal point of the mirror, (c) side view on another mirror tip to assess how far the tip protrudes, (d) yet another tip where the writing process failed, (e) surface at the outside of the paraboloid coated with a thin metal layer. In the top left one can see the scaffold which makes up the volume of the wall.

The sharp apex of the STM tip proved difficult to reproduce, though, as this extraordinarily small feature size pushes the technological limit of the DLW setup. As a result, many of the fabricated tips had a much larger tip radius, see Fig. 3.2(c), or sometimes did not even resemble a tip at all. An example of the latter is shown in Fig. 3.2(d). Interestingly, this tip still yielded a decent electroluminescence signal and was used for the spatially resolved photon map discussed in Sec. 4.2.3, even though it featured a very dense spectrum with numerous overlapping peaks, which is typical for blunt tips.

Additionally, it proved difficult to place the STM tip precisely at the focal point of the parabolic mirror. In Fig. 3.2(c), a mirror tip was imaged from its side and it was estimated how far the tip apex protrudes from the mirror edge in the foreground. This protrusion was designed to be 1.4 µm, but in the SEM picture it amounts to approximately 3 µm. This might be caused by a slight distortion of the strut due to the shrinkage during the development process. However, this does not pose a serious problem, as this deviation would — assuming geometric optics and a perfect mirror shape — only cause a beam divergence of 5°, which is inside the acceptance angle of the fibre. Also the impact on the covered solid angle due to the larger opening at the side is limited, and only amounts to approximately 10 %. Currently, the DLW procedure is being further optimised to increase the reproducibility of the tip radius and position.

The smoothness of the mirror surface is difficult to assess, for it is inaccessible for any surface characterisation technique. Vicariously, the outside of the mirror shape — written with the optimised slicing distance of 200 nm — is presented in Fig. 3.2(e) as measured by SEM. The ripples from the individual writing layers are still faintly visible, however, the absolute value of the roughness cannot be measured in our SEM. Instead, the DLW structures reported in [185] are referred to as an example. They were written with the same resist at similar settings and their surface roughness was determined to be around 5 nm using atomic force microscopy. This, however, only accounts for the roughness of the DLW structure. The roughness of the thick metal coating on the reflective side of the mirror will add to this, and the total roughness σ is therefore expected to be larger than for conventional mirror surfaces. This will result in a mixture of specular and diffuse reflection. According to Bennett and Porteus [186], the fraction of the specular reflection R_S relative to the total reflected intensity R_0 in the long-wavelength limit amounts to

$$\frac{R_S}{R_0} = \exp\left(-\frac{(4\pi\sigma)^2}{\lambda^2}\right). \tag{3.1}$$

If, as an example, the surface roughness is $\sigma = 15\,\text{nm}$, the specular reflection at a wavelength of 500 nm amounts to 87 %, not including material absorptivity. The corresponding loss is not too concerning, but neither is it negligible, so it will be grounds to further optimise the mirror tips in the future. One possibility to decrease the roughness of the metal coating in the future is to perform the vapour-deposition while the mirror tips are at low temperature, which prevents the formation of large grains by suppressing surface diffusion and could be done in a setup of Prof. apl. Detlef Beckmann at the *Institute of Nanotechnology*. However, not all of the diffusely reflected light is lost, as the optical fibre will transmit at least a fraction of it, which happens to hit the fibre end within its acceptance angle. Here, the advantage of having the fibre directly attached to the collection optics pays off.

3.1.2 Optical fibre

A multi-mode step-index silica fibre (*Thorlabs* UM22-200) with a core diameter of 200 µm, a cladding diameter of 220 µm and a UHV compatible polyimide jacket transmits the light from the mirror tip to the detector. The fibre has a larger diameter than the mirror to account for potential misalignments. The fibre is fed through a small cable duct in the liquid helium cryostat and holes in the triple-stage radiation shields up to a UHV feed-through at the top end of the cryostat. Special care was taken not to create a direct optical pathway for heat radiation to reach the STM by choosing holes that are offset against each other and by additionally covering the holes with aluminium foil. To avoid intensity loss in the optical fibre due to additional interfaces at the UHV feed-through, a custom feed-through was built in which the fibre is not interrupted. The fibre is glued inside a commercial 1.25 mm ceramic ferrule which itself is glued inside an equally sized bore in a blind flange using *EPO-TEK* 353ND adhesive. When it is applied, this adhesive is very inviscid and fills all gaps between the components, creating a flawless seal.

The total length of the optical fibre is ~5.5 m of which 1.3 m is located inside the UHV chamber. The section inside the UHV chamber was covered with a fibre glass protective tube, so that the fibre cannot rub against any edges. The section outside the chamber was covered with a light-tight polymer tubing and inserted into a semi-flexible stainless steel tubing, which protects the fibre from accidental bending. The fibre was terminated on the ambient side with a standard SMA connector.

It was experimentally verified that low temperatures, high temperature gradients or thermal shocks do not cause unwanted attenuation within the optical fibre, as it will be exposed to these conditions within the cryostat. For that purpose, a laser diode was focussed into a test fibre and a photo diode connected to a multimeter monitored the light through-put while the fibre was repeatedly immersed in liquid nitrogen and warmed up to room temperature again. The same technique was also used to check how much a tight radius in the fibre degrades the transmission, as the fibre has to take a sharp turn after exiting the STM to fit through a cable duct of the cryostat. The bending radius at this spot is 30 mm for two subsequent quarter turns in the worst case, which is below the minimum long-term bending radius of 53 mm, as rated by the manufacturer. However, it was found that one full revolution at such a low bending radius results in a transmission reduction of merely 3.5 %. Therefore, roughly half that value should apply to the two subsequent quarter turns, which is acceptable.

3.1.3 Spectrometer and detector

The collected light from the tunnelling junction is analysed by a commercial *Acton SP-2156* imaging spectrograph equipped with a 300 grooves per millimetre diffraction grating combined with a *PyLoN 100B* liquid nitrogen cooled charge-coupled device (CCD) camera by *Princeton Instruments*. This spectrograph implements the Czerny-Turner design [187], and its working principle is shown in Fig. 3.3. Incoming light needs to be focussed onto the entrance slit of the spectrograph, which is usually set to the width of the focus. The light beam is redirected by the flat mirror M1, collimated by the parabolic mirror M2 and directed onto the diffraction grating. The grating features a reflective, sawtooth-like, periodic surface structure, which is designed to divert up to 70 % of the incoming light into the first order of diffraction. Different wavelength components of the light are then separated into different outgoing angles and then refocussed by a second parabolic mirror M3 onto the imaging sensor of a CCD camera. By that, images of the entrance slit are reproduced on the sensor at different locations for different wavelengths.

The sensor features an array of 1340 × 100 square pixels, each of them 20 μm wide, in which photons are absorbed. Hereby, electron-hole pairs are created, which are immediately separated. During the exposure time, the generated charges accumulate in the so-called well of each pixel. After exposure, the content of each well has to be read. By applying voltages to gate electrodes at the wells in a specific order, these charges are subsequently shifted through the array into a read-out register at the input of a low-noise charge amplifier. This way, each pixel can be addressed individually, but it is moreover also possible to shift the charges of several pixels into the same register before reading its content. By that, the charges originating from different pixels can be added, and then their sum is read out only once (*binning*). This is very useful for obtaining spectra, because only the horizontal intensity distribution at the CCD contains information about the wavelength composition of the incoming light, and the vertical distribution is irrelevant if only one light source is involved. Therefore, all vertical columns are typically added if a spectrum is acquired, which reduces the two-dimensional image of the camera to a one-dimensional spectrum. This is much faster than reading each and every pixel subsequently, and by adding the charges by hardware, the read-out noise of the charge amplifier applies only once per column, and not for each of the 100 pixels per column individually. Therefore, hardware-based binning is also useful for detecting weak signals, which yield only few counts per column. By adding the charge contents of several adjacent columns (*horizontal binning*), the photon counts per bin are even more increased at the expense of reduced wavelength resolution. This technique was used for all spectra presented in this work.

The CCD camera is liquid nitrogen cooled and operates at −120 °C. This is necessary because at ambient temperatures, thermally induced charge separation in the semiconductor pixels causes a so-called dark current, which easily exceeds the light induced charge separation for weak sources like STML. At the operating temperature, this dark current is decreased to one thermally excited electron per 12.5 h and pixel, as tested and specified by the manufacturer. For exposure times of a few seconds, this is entirely negligible even if many pixels are binned together.

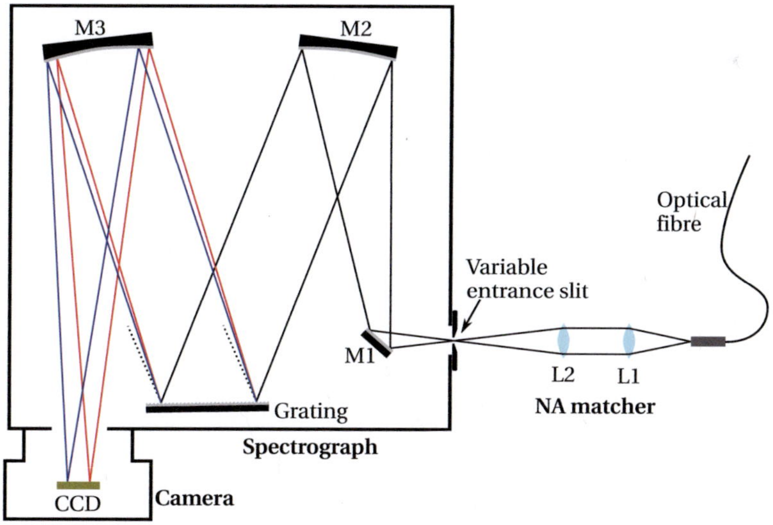

Figure 3.3: Principle of the implemented light detection setup. The light cone from the optical fibre passes a NA matcher, is coupled into a Czerny-Turner style spectrograph with a reflective diffraction grating and detected by a liquid nitrogen cooled CCD array featuring 1340 × 100 pixels. The dotted line originating from the grating indicates the zero-order reflection. The purpose of the mirrors M1 to M3 and the lenses L1 and L2 is explained in the text.

From Fig. 3.3 it is apparent that the size of the mirrors inside the spectrograph limits the maximum tolerated divergence of the incoming light cone. This is specified as the NA of the spectrograph, which is easily translated into an acceptance angle. For the spectrograph installed in this setup, this NA is 0.12, corresponding to an acceptance angle of 6.9° relative to the beam axis. However, the optical fibre which transmits the light signal to the analyser has a NA of 0.22, meaning that the opening angle exiting the fibre is higher than the spectrometer accepts. To fix this mismatch, two achromatic lenses with a respective focal length of 25 and 50 mm (*Edmund Optics* 49323 and 49328) were added between the fibre exit and the entrance slit of the spectrograph. These lenses correspond to L1 and L2 of the *NA matcher* in Fig. 3.3. They magnify the image of the fibre end end by a factor of two onto the entrance slit of the spectrograph, thereby narrowing the opening angle of the light cone by the same factor.

Compared to coupling the fibre directly to the entrance slit, the slit now has to be opened to 400 μm instead of 200 μm to let all the light pass through. This large spot is then reproduced at the CCD sensor of the camera by the optics of the spectrograph, which means that even monochromatic light will be dispersed across a diameter of 400 μm on the CCD, covering a circle of 20 pixels in diameter. The wavelength resolution of the spectrometer setup is therefore not determined by the pixel size of the camera, but by the size of the focus at the entrance slit, and amounts to 8 nm. As a consequence of this wide broadening across many pixels, the aforementioned horizontal binning does not have a negative impact on the wavelength resolution, and was chosen to result in 1.6 to 4 nm wide bins throughout the measurements in this work. In the future, however, it will still be possible to sacrifice signal intensity for higher resolution by reducing the width of the entrance slit whenever this seems appropriate.

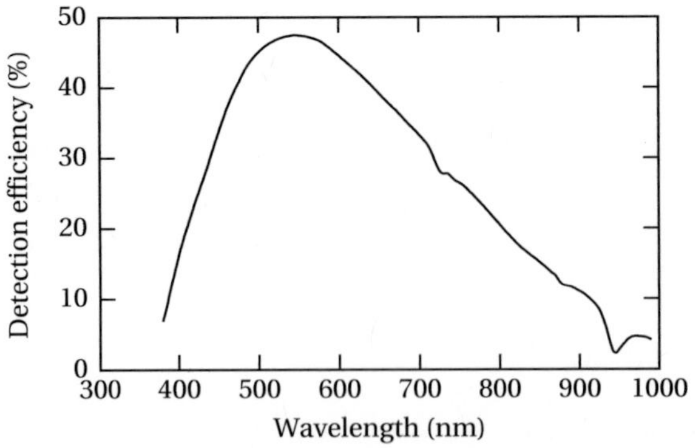

Figure 3.4: Theoretical detection efficiency of the light path considering all optical components from the fibre entrance at the STM to the CCD.

It is noteworthy that it is necessary to focus the additional lenses at the spectrograph input, which seems to contradict earlier statements in this work claiming that no focussing is required in this design. However, these two lenses had to be focussed only once, and it could be done while replacing the weak STML source with a bright light source, which makes the process much easier. In other designs, the whole setup has to be refocussed after each tip change, and even more often if the focus inside the STM changes due to thermal drift. Also, the focussing procedure in other STML setups is cumbersome because it relies on the extremely weak light signal of the STM tip itself. None of that is the case in this design.

The light exiting the mirror tip inevitably suffers from intensity loss at the fibre interfaces, inside the fibre, at the NA matching lenses and mirrors of the spectrograph. Considering these losses combined with the diffraction grating efficiency and finally the conversion efficiency of the CCD camera, the wavelength response of the optical setup amounts to the curve in Fig. 3.4, showing that light can be detected in the wavelength range from 400 to 1000 nm. As the actual efficiency of the mirror tips is still largely unknown, the response curve in Fig. 3.4 does not account for this, but rather represents the upper limit of the possible total detection efficiency, which may be further approached in the future by optimising the writing procedure and the metal coating of the mirror tips.

3.2 Scanning tunnelling microscope

The home-made STM shown in Fig. 3.5 was designed according to the Pan-style [101], and, in some aspects, similar to previous STMs made in our group [103]. The placement of the piezo scanner tube and the tip relative to a single-crystal sample is evident from Fig. 3.5(b). The tip is located above the sample because the optical fibre is fed through the cryostat from the top and cannot bend very much, therefore a sharp turn on its way to the scan head had to be avoided. At its maximum, the STM body has a diameter of 46 mm and a height of 76 mm. The inner-most radiation shield of the cryostat has at its narrowest a diameter of 55 mm, so the STM easily fits inside.

3.2.1 Materials

The body of the STM needs to be stiff and thermally conductive to provide low STM noise and facilitate quick thermalisation during system cool-downs and after sample transfers. Tab. 3.1 lists the thermal conductivity of some popular materials for

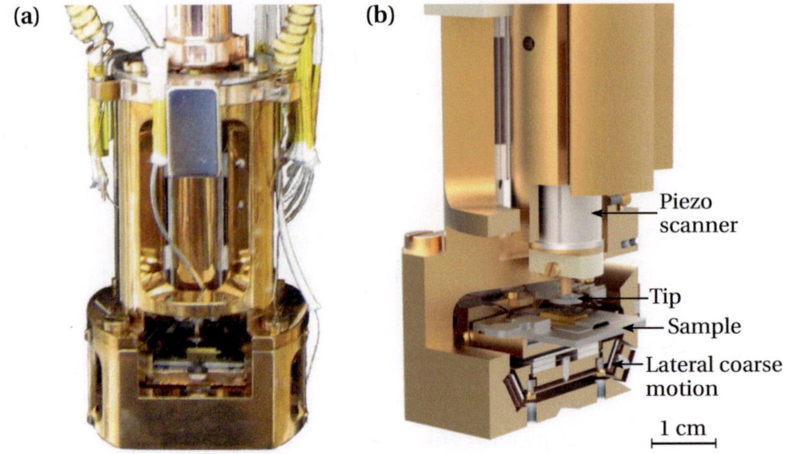

Figure 3.5: The γ-STM. (a) Photograph of the finished STM, (b) sectional view of the design indicating some key components.

building cryogenic devices. Copper provides by far the highest thermal conductivity among these, but it is rather soft. So age hardened beryllium copper (material number 2.1247) was chosen for the STM body instead, as it is much tougher, has a 10 % higher elastic modulus and still provides acceptable thermal conductivity compared to other options.

Insulating components were manufactured either from Macor or from polyether ether ketone (PEEK). The first is a machinable, artificial, glass-like ceramic, the latter is one of the few UHV-compatible plastics, and quite tough at that. Parts which feature internal threads or are held in place by screws were made from PEEK because Macor is extremely brittle.

The scanning action of the STM tip is performed by a 30 mm long and 10 mm wide segmented lead zirconate titanate (PZT) piezo tube (*Physik Instrumente* PT230.24). The $5 \times 5 \times 0.75\,\mathrm{mm}^3$ shear piezo plates that constitute the slip-stick motor (see below) are also made from PZT that is polarised parallel to the electrodes. For the sliding surfaces of the stick-slip motor, polished alumina and sapphire were chosen, mainly because this has worked well in previous STMs that have been built in our group.

Table 3.1: Heat conductivity of popular materials in cryogenic devices at 4 K in $\mathrm{W\,m^{-1}\,K^{-1}}$. Values taken from [188, 189] except for titanium [190] and Macor [191].

Copper	Beryllium copper	Stainless steel	Titanium	Macor
300 to 1000	1.9	0.3	0.05	0.08

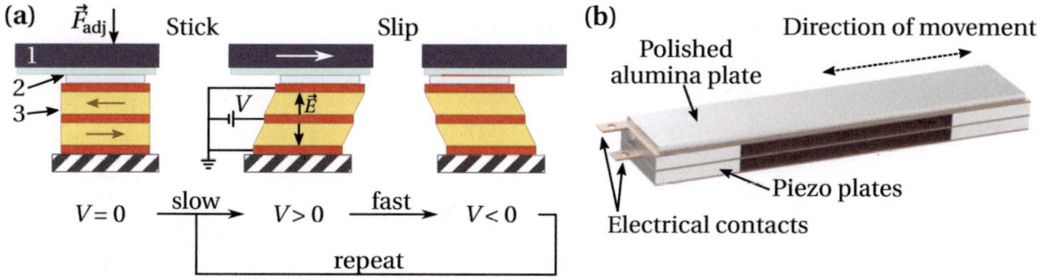

Figure 3.6: Operation principle of the coarse motion. (a) Illustration of the stick-slip principle with (1) moving stage, (2) sliding interface and (3) actuator consisting of two shear piezo plates with opposing polarisation directions (indicated by brown arrows). The piezo voltage V is applied to the middle electrode with respect to the outer ones. An adjusting force F_{adj} determines the friction of the sliding interface. (b) Design of the actuators used in the γ-STM. Two stacks, each consisting of two piezo plates, support a mutual sliding interface and copper sheets act as electrical contacts.

All metal-ceramic interfaces were glued with *Epotek* H77S and H20E epoxy glue for electrically insulating and conductive interfaces, respectively. The glue requires curing at 150 to 180 °C, which, if done in air, would oxidise the metal surfaces and thus compromise subsequent bonding processes, as the glue only adheres well to bare metal. Curing the epoxy was therefore done in a separate vacuum chamber at approximately 1×10^{-2} mbar.

3.2.2 STM body and coarse motion design

The scanner tube provides only a very limited moving range for the tip of 1 µm laterally and 150 nm vertically. To exchange either the tip or the sample, position them with respect to each other at a millimetre scale and finally approach the tip to the range of the scanner tube, another positioning system with a much higher range was integrated into the STM, which is referred to as *coarse motion*. In this, the sample is moved laterally and the tip assembly including the scanner tube vertically, utilising the popular inertial slider or stick-slip principle [102]. This is compatible with UHV and low temperatures and moves in very small steps, which is necessary to approach the tip into the range of the scanner tube without overshooting and crashing the tip into the sample. Once the tip is in tunnelling contact, the coarse motion is switched off and then purely relies on static friction to hold the stages in place, without requiring any holding current or voltage. Therefore, no energy is dissipated during the measurement, which is essential to reach low temperatures, and the sample and the tip are hold at their respective position without additional vibration caused by the motor.

The principle of operation is illustrated in Fig. 3.6(a) and relies on shear piezo plates that exhibit a tiny dislocation perpendicular to the applied electric field. The moving

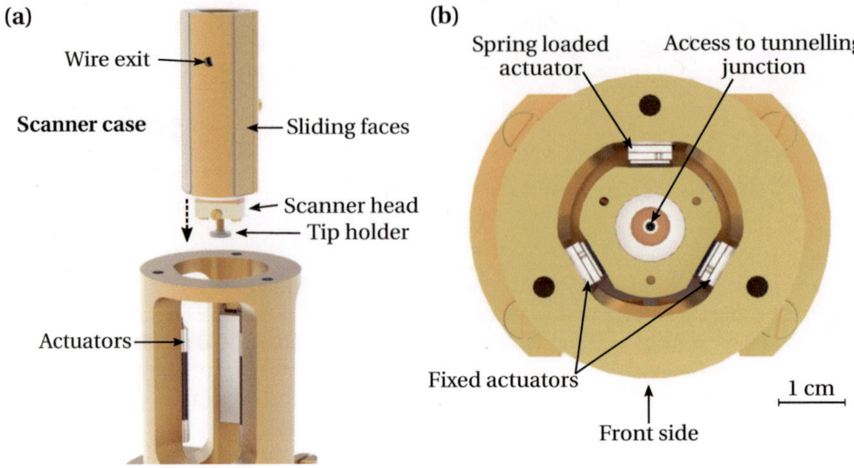

Figure 3.7: Vertical movement of the scanner case inside the STM body. The cap of the STM, which attaches to the suspension springs and holds the permanent magnets, was removed for clarity. (a) Scanner case slid out of the body, (b) top view of the scanner case mounted by the three sliding interfaces. In all components, centre holes clear the path of the optical fibre, which is attached directly above the tunnelling junction (see later).

stage is supported by these piezo actuators with a sliding interface in between, and the piezo actuators cycle through slow and fast back-and-fourth displacements. During the slow cycle, the friction between actuator and the moving stage is sufficient to move the latter to the side (*stick*). Subsequently, the actuators move back as fast as possible, so that the friction is overcome (*slip*) because the inertia of the moving stage prevents it from following the strong acceleration of the actuators. Then, the stage has advanced by one step and the actuator is ready to start another cycle. Repeating this with a frequency of 1 to 10 kHz results in a net movement in the order of 1 mm s^{-1}. For this to work, the friction at the sliding interface needs to be finely adjusted to a specific value, though. If it is too tight or too loose, either the slip or the stick cycle are not going to work. Adjusting the friction is achieved by setting a variable clamping force F_{adj} to a critical range, which has to be found by try and error.

The piezo actuators used in the γ-STM are depicted in Fig. 3.6(b). In each actuator, two shear piezo plates are stacked on top of each other to increase the total travel. As indicated by the brown arrows in Fig. 3.6(a), the polarisation directions of both plates oppose each other, so that the piezo voltage only needs to be applied to the middle electrode, and the outer ones are grounded. Then, the electric field inside the top plate is opposing the field in the bottom plate, which, considering their opposing polarisations, results in a movement into the same direction. In fact, the grounded STM body itself comprises the lower electrode because the stacks are glued directly to the body

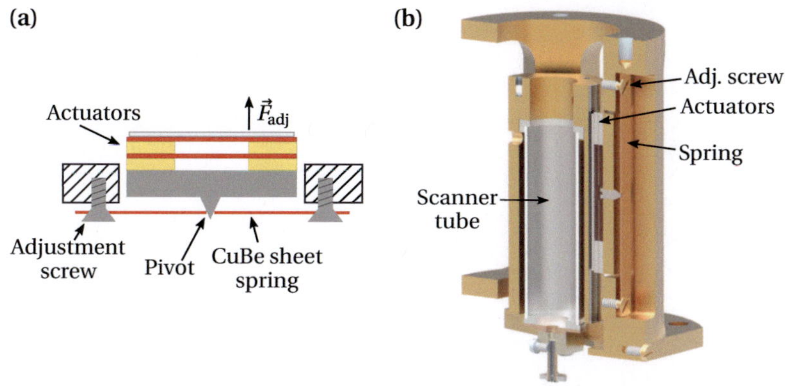

Figure 3.8: Spring loading of one actuator per moving stage, (a) rough sketch adapted from [101], (b) sectional view of the implementation for the motor in the vertical direction at the back of the STM body.

with no insulation in between. To prevent the moving stages from rocking, each sliding interface is supported by two stacks of piezo plates in parallel, which are located as far apart as possible.

In the following, it will be detailed how exactly each of the three individual coarse motion stages are assembled and mounted. In Pan's design [101], the scanner holder is mounted between three sliding planes oriented at 120° with respect to each other, two of which are fixed and one is spring loaded with two adjustment screws for the spring tension of each stage. The orientation of the sliding planes relative to each other ensures that the adjustment force is equally distributed among them and, thus, the friction is the same for all sliding interfaces. This idea is reproduced in the design of the γ-STM, as can be seen in Fig. 3.7. The scanner tube is mounted inside the scanner case, which moves vertically within the STM body. In Fig. 3.7(a), the scanner case was slid out to the top to expose the sapphire sliding faces of the scanner case and the actuators mounted within the STM body. Once the scanner case is pushed into the STM body, Fig. 3.7(b) shows how the sliding faces meet and prevent the scanner case from moving in any direction except the vertical.

The spring loading mechanism needs to be designed such that the actuators are rigidly fixed in the direction of movement. Otherwise, the adjustment spring would allow the spring loaded actuator to move during the stick phase, instead of imposing its displacement onto the moving stage. Also, the adjustment force needs to be evenly distributed across the sliding interface. The solution of Pan *et al.* to this problem is sketched in Fig. 3.8(a). The spring loaded actuator is supported by a metal sheet spring that is fixed along the direction of movement with two countersunk adjustment screws. The adjustment force is transferred to the actuators at a pivot in the middle of the sheet

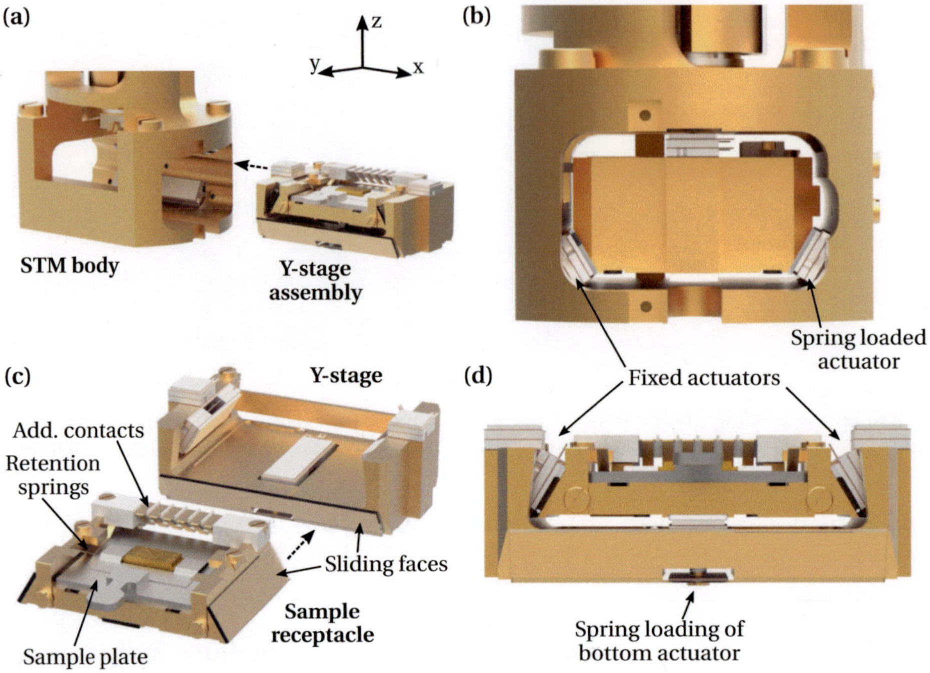

Figure 3.9: Design of the nested stages responsible for moving the sample receptacle in the lateral X and Y dimensions. (a) Exploded view of the Y stage assembly slid out of the STM body. (b) Side view of the Y-stage assembly inside the STM body. The end stop was omitted for clarity. (c) Sample receptacle slid out of the Y-stage and (d) sample receptacle in its installed position inside the Y-stage.

spring to evenly distribute it even if the screws are not tightened equally. Otherwise, the adjustment force would vary depending on the position of the moving stage. In Fig. 3.8(b), the implementation of this mechanism in the γ-STM is demonstrated for the vertical positioning stage. The implementation of the spring loading of the other two stages is similar.

In previous STMs made in our group [103], lateral positioning stages are based on a different mounting geometry. However, the Pan-geometry has proved the most reliable, so the lateral moving stages of the γ-STM were designed in a similar fashion, which is shown in Fig. 3.9. The coordinate system in Fig. 3.9(a) clarifies the designation of the axes which are referred to in the following. Two-dimensional movement of the sample is achieved by nesting two stages. The sample plate is inserted into the *sample receptacle* — seen best in Fig. 3.9(c) — which is connected to the bias voltage and fixes the sample plate by two retention springs to prevent it from rocking. To accommodate microstructured samples or sensors mounted on the sample plate in the future, the sample receptacle also provides six additional sample contacts made from beryllium copper springs. As can be seen in Fig. 3.9(c), it slides into the intermediate Y-stage, which contains the actuators to move the receptacle in the Y direction. Fig. 3.9(d) shows how the sliding interfaces within this stage line up. The assembly of the Y-stage and the sample receptacle then slides into the STM body, which is shown in Fig. 3.9(a). Inside the STM body, two actuators for the X direction are located at the bottom. The top actuators were not mounted within the STM body, but on the Y-stage itself. They are polarised into the opposite direction compared to the bottom actuators to achieve a net movement of the Y-stage into the same direction. This was necessary to make the design as compact as possible. The arrangement of all actuators responsible for the movement in the X direction is presented in Fig. 3.9(b). To provide some flexibility regarding the investigated samples and deployed tips, the coarse motion was designed for a wide range of 8 mm × 4 mm in the lateral directions and 12 mm for the vertical tip displacement.

Reliable coarse motion operation is only achieved if the clamping force acting on the sliding interfaces stays constant regardless of the current position of the moving stage. As a consequence, the sliding planes, which confine the moving stages, have to be as parallel along the travel direction as possible. Therefore, special care was taken to fabricate the actuators to a very consistent height by applying even pressure during the glueing process in a home-made jig which utilises two equal masses to apply even pressure at two points. It was verified with a micrometre screw that the height difference between two ends of each actuator was in the order of 10 μm.

3.2.3 Mirror tip holder

The tip holder has to provide a mean to grab the tip securely for transferring it through the UHV chamber, and it also has to provide vertical and lateral alignment of the mirror tip relative to the optical fibre end once it is installed in the STM. In design of the γ-STM, axial alignment is provided by an optical ferrule and a mating sleeve. These components are commonly used to temporarily mate two fibre ends and are commercially available. A mating sleeve is shaped like a tube with a slit cut into the side, that opens up a little once it is pushed onto the round ferrule, which houses the fibre precisely at its centre. When these components are put together, the axis of the mating sleeve very well coincides with the axis of the ferrule due to their precise fabrication. In this design, the optical fibre is glued into a 1.25 mm stainless steel ferrule (*Thorlabs* SFLC230-10), which is fixed at the bottom end of the scanner tube of the STM, as will be detailed in Sec. 3.2.4. The changeable tip holder is in turn equipped with a matching phosphor bronze mating sleeve (*Senco* PBS-12-680), which is shortened to 4.4 mm

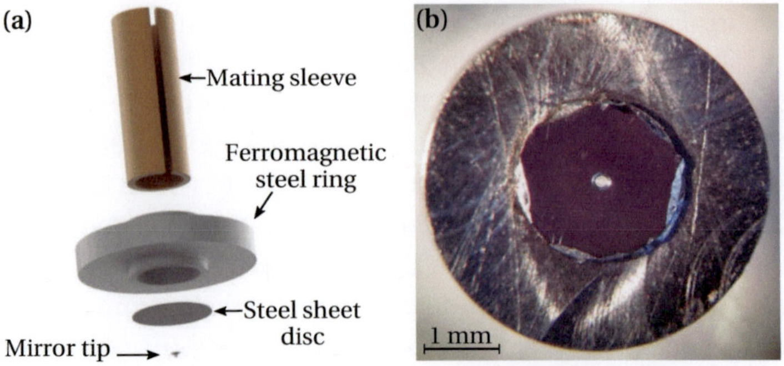

Figure 3.10: Holder used to handle and align the microscopic mirror tip inside the STM. (a) Exploded view of the assembly. (b) Photograph taken from the bottom side showing the still uncoated mirror tip in the centre of the polished steel substrate.

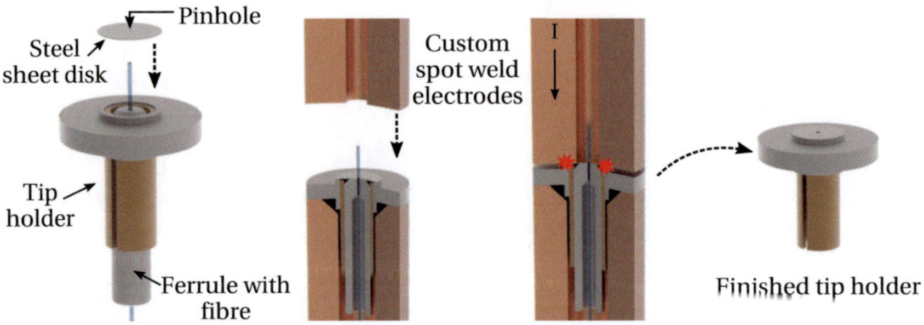

Figure 3.11: Spot welding the steel sheet disc onto the tip holder such that the pinhole is centred with regard to the mating sleeve.

and glued to a ferromagnetic steel ring, see items 1 and 2 in Fig. 3.10(a). The reason to favour a metallic ferrule and mating sleeve over the much more easily available ceramic ones lies in that these components have to conduct the tunnelling current.

For the vertical alignment of the mirror tip, which will be explained in the next section, all used tip holders have to be of precisely the same length, which seems difficult at first as the mating sleeves are shortened manually. For that purpose, an assembly procedure was conceived which achieved a length consistency among different tip holders of less than 10 μm. This is possible by applying epoxy glue to the the mating sleeve and pushing it into the steel ring between the anvil and the spindle of a micrometre screw. Once the desired length is reached, as indicated by the digital readout of the micrometre screw, the tip holder is released and the viscosity of the yet uncured glue is sufficient to keep the two components in place for the subsequent curing process.

The aforementioned 25 μm steel sheet disk is spot welded onto the bottom end of the steel ring. For the pinhole in the steel sheet disk to coincide with the axis of the mating

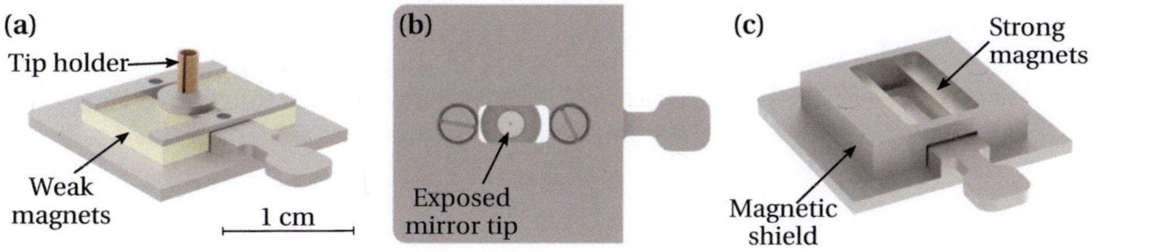

Figure 3.12: Carrier plates used to handle the mirror tip in UHV and transfer them (a),(b) into and (c) out of the STM. In (b) the plate is depicted from below and the exposed mirror tip can be coated from this side using vapour-deposition.

sleeve, it has to be properly centred during spot welding. The procedure that was conceived for this purpose is illustrated in Fig. 3.11. The alignment was ensured by temporarily inserting another ferrule into the mating sleeve with a protruding 130 μm optical fibre. The fibre pokes through the pinhole and fixes it concentrically to the mating sleeve. Then, the steel sheet disk can be spot welded between two custom copper electrodes. Thereafter, the front end of the finished tip holder is polished using a custom jig which ensures that the fibre end is pressed flat against the polishing sheets without introducing an undesired angle. Hereby, also the burrs from both the laser cutting and the spot welding are removed.

For the subsequent DLW process, the pinhole is used to identify the optimal placement for the mirror tip, which is then written on top of it. A micrograph of the finished and yet uncoated assembly is presented in Fig. 3.10(b). It shows the DLW mirror tip in the centre of the polished substrate on top of the, now obscured, pinhole.

The tip holder assembly is handled in UHV with a carrier plate that engages the steel ring of the tip holder with two weak magnets, suspending the delicate mirror tip in between. This carrier plate is also used to hold the tip holder during the *in-situ* metallisation process. For the metal vapour beam to reach the mirror tip, a hole in the carrier plate exposes it from the bottom. Figs. 3.12(a) and (b) show the carrier plate used for coating the mirror tips and inserting them into the STM from the front and bottom. In Fig. 3.12(c), a similar carrier plate is depicted which, however, houses two stronger magnets. These are necessary to remove a tip holder from the STM again, see the following section.

3.2.4 Tip alignment and exchange

The scanner head at the bottom of the scanner tube connects the changeable tip holder to the STM. As shown in Fig. 3.13, it features a grounded guard electrode to prevent

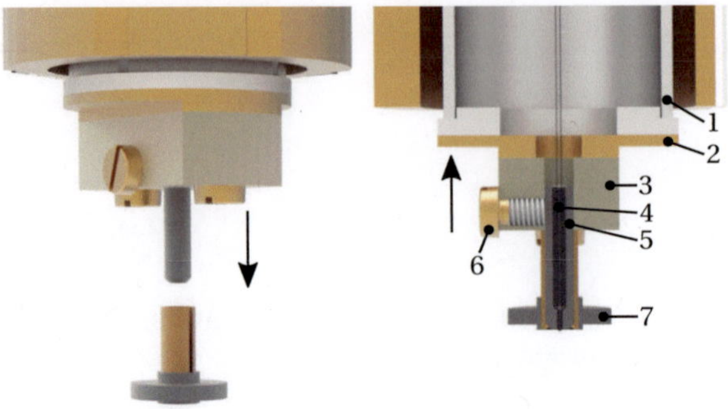

Figure 3.13: Installation procedure for a new mirror tip. The scanner head picks the tip holder from a weak magnetic carrier plate (not pictured). Once the ferrule is fully inserted into the mating sleeve of the tip holder, the mirror tip is guaranteed to be axially aligned with the optical fibre and as close to it as possible. (1) Scanner tube, (2) grounded guard, (3) insulating PEEK ferrule receptacle, (4) optical fibre, (5) ferrule, (6) setscrew, (7) tip holder.

electronic cross-talk between the piezo drive signals and the tunnelling current, that is insulated from the scanner tube by a Macor ring. Onto the grounded guard, a block of PEEK is screwed, which holds the ferrule. A setscrew in the PEEK block locks the ferrule and additionally acts as an electrical contact for the tunnelling current cable. The length that the ferrule protrudes from the bottom of the PEEK receptacle is set to the same length of the tip holder, such that the mating sleeve touches the PEEK block almost at the same time as the steel sheet disk is touching the polished fibre end of the ferrule. This is the reason why all used tip holders have to have precisely the same length, and ensures that the steel sheet disk is as close to the fibre end as possible, but protected from being pushed too far against the ferrule and bent.

The optical fibre is glued into the ferrule, fed through the inside of the scanner tube and exits the STM to the top. It is not rigidly fixed anywhere in the cryostat and, thus, flexible enough to tolerate the vertical movements of the scanner tube assembly during tip and sample changes.

To install a new tip on the scanner head, the tip holder sitting on a carrier plate with a weak magnet is placed underneath the retracted scan head, as illustrated on the left of Fig. 3.13. The lateral coarse motion is then used to visually align the tip holder with the ferrule with the help of a magnifying camera installed on the air side of a UHV viewport. Then, the scanner head is lowered and the ferrule slides into the mating sleeve of the tip holder. When the ferrule is fully inserted into the mating sleeve, the clamping force is high enough to pick up the sample holder from the magnetic carrier plate when the scanner head is raised again. As the optical fibre is concentric with the ferrule and the mirror tip is centred with regard to the mating sleeve, the mirror tip is now also axially aligned with the optical fibre. To remove the tip holder again, another carrier plate with stronger magnets is placed underneath and the scanner head is slowly lowered until the magnets pull the tip holder from the ferrule.

To permit this procedure in UHV, the ferrule had to be modified. The most obvious problem that would otherwise occur is the excessive force required to insert the original ferrule into the mating sleeve, which is designed to amount to 40 N in ambient conditions. This force would exceed the strength of the coarse motion by far and also put unnecessary stress on all components, especially because friction is considerably higher in UHV. Thus, the clamping force that the mating sleeve exerts on the ferrule must be reduced. When sliding the ferrule into the mating sleeve, it is also not favourable to have the mating sleeve fully engage the ferrule right from the beginning, as it might cant and get stuck before it is fully inserted. Finally, one needs to account for the fact that the mating sleeve is glued into the steel ring of the tip holder at its bottom end and is, therefore, not capable of bending there.

To remedy these issues, the shape of the ferrule was carefully modified and consists of two sections of different diameters. The lower section, which spans about 4 mm from the fibre end upwards, was thinned down enough to be barely thinner than the inner diameter of the disengaged mating sleeve. Thus, this section will slide into the mating sleeve without clamping force and fits all the way down to the bottom end of the tip holder. Reducing the diameter of the ferrule was done manually with sandpaper while the ferrule was spinning, so there is no sharp transition between the lower and upper section but rather a short taper. That prevents the ferrule from getting stuck while sliding down. Only when the ferrule is almost fully inserted into the tip holder, the mating sleeve engages the ferrule at its thicker upper section. The length at which the sleeve holds on the ferrule in the end is much less than a millimetre. The upper section was only carefully thinned to reduce the clamping force down to a point where it is strong enough to pull the tip holder from the weak magnets of the carrier plate, but weak enough to let it go once the strong magnets are placed underneath.

These modifications of the ferrule will compromise the exact alignment of the mating sleeve, however. This is the main reason why an optical fibre was chosen which is larger than the mirror opening.

3.2.5 Mechanical decoupling

To decouple the STM from vibrations of the cryostat, especially caused by the bubbling of the liquid nitrogen and incomplete decoupling of the chamber from the environment, a classic passive approach was followed, a combination of a damped suspension and a very rigid STM body [192]. During measurements, the STM is suspended by three phosphor bronze springs. This constitutes a low-frequency oscillator with a resonance frequency of around 3 Hz. It is essential to provide dampening to this oscillator to achieve low vibration transfer. To that purpose, three NdFeB permanent magnets were added at the top of the STM body. These induce eddy currents in the surrounding radiation shield upon vertical movement. Two additional permanent magnets are located directly below the STM at the bottom of the radiation shield to dampen lateral movements as well. The permanent magnets have a 120 °C rating to facilitate baking of the system. Additional dampening was achieved by inserting polyimide foam into the phosphor bronze springs.

The STM body itself was designed as rigid and compact as possible and, therefore, constitutes a high-pass filter. In an attempt to characterise the rigidity of the STM, a finite element simulation was performed to determine its eigenmodes of vibration. The frequency of the lowest mode which modifies the tip-sample distance determines the vibration transfer to the tunnelling junction and should be as high as possible. The actual drawings of the STM in full detail are, however, impossible to simulate on an office computer with the built-in module of the CAD software due to their high complexity. Therefore, many details like holes, threads, minor rounded edges and the thin sheet springs had to be omitted or modified until the simulation succeeded. The most important limitation of the utilised software is, however, the lack of support for assemblies that consist of different materials. Therefore, the STM including all ceramic components was modelled as one solid block of beryllium copper, whose elastic modulus was specified in the software to be 130 GPa [193]. This is a serious simplification, not only because the elastic modulus of the ceramic components is different, but also because the glued and sliding interfaces were neglected that way. In particular, epoxy glued interfaces are known to significantly decrease the resonance frequency of a system, compared to finite element simulations in which the adherent is neglected [194].

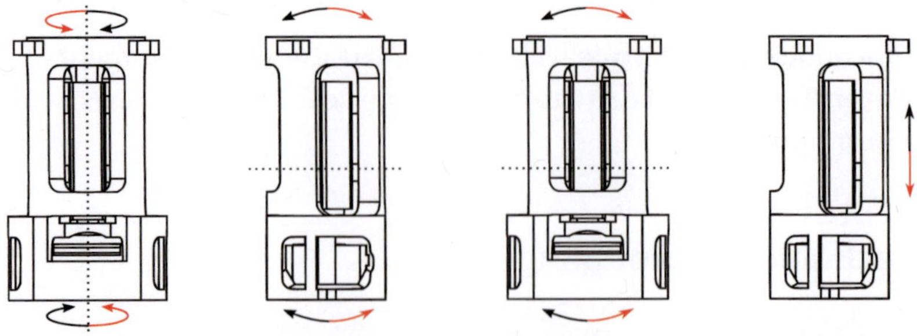

Torsion mode 3930 Hz — Tilt mode 4860 Hz — Tilt mode 6130 Hz — Axial mode 7640 Hz

Figure 3.14: Lowest modes of vibration of the assembled STM body as found by a finite element simulation. The arrows and their colour indicate the direction and relative phase of movement, the dotted lines indicate the nodes at which the material is almost stationary.

As a consequence, the results obtained with this procedure are to be treated with caution. However, this still proved a useful tool to optimise the STM body during the design phase, because the influence of design modifications on the resonance frequency could be qualitatively assessed. In Fig. 3.14, the lowest eigenmodes of the final STM body design are presented and categorised. The torsion mode, which has the lowest frequency of them, does approximately not affect the tip-sample distance and can be neglected. Therefore, the noise-rejection performance of the STM is determined by the first tilt mode at ~4.9 kHz.

Another important strategy to reduce mechanical noise in the tunnelling junction is to avoid vibrations of the cryostat and the UHV setup in the first place. Therefore, all the bell-shaped radiation shields and the suspended liquid helium tank were damped with polyimide foam. Also, all long and protruding UHV components on the machine were properly supported against the frame at both ends.

3.2.6 Instrumentation

In the γ-STM, the sample receptacle is connected to the bias voltage and the tip is virtually grounded by the input of a *Femto* DLPCA-200 transimpedance amplifier, which is directly attached to the UHV feed-through and measures the tunnelling current. The voltage signals for the scanner tube, coarse motion and sample bias are supplied by a commercial *Nanonis* SPM controller, which also measures the output of the transimpedance amplifier and implements the feedback loop. It also features a built-in lock-in amplifier for spectroscopy. The *Nanonis* software provides a *LabView* interface, as well as the *Lightfield* controller software of the optical spectrometer and CCD camera. This facilitates fully automatic long-term experiments in which the tip position, bias voltage and current setpoint can be freely varied to study their influence on the light emission. For the experiments presented in this thesis, I have programmed a versatile *LabView* automation library, which will also be useful for future experiments.

Using a four-probe technique, the temperature of the STM is determined by measuring the resistance of a *Lakeshore* Cernox CX-1010-SD-HT sensor, which is screwed onto the back of the STM body. The temperature dependence of this sensor was calibrated by Arnold Seiler at the *Physikalisches Institut* against well calibrated thermometers with an estimated accuracy error of less than 5 % between 1 and 4 K.

3.3 Cryostat

Many STML setups operate at room temperature, as the temperature has little effect on the plasmonic luminescence of metal surfaces, and working at low temperatures involves serious complications of the entire experiment. The γ-STM is, however, specifically designed with single molecule studies in mind, and molecules often exhibit thermal diffusion on noble metal surfaces at room temperature [195], which constitutes a problem for inherently slow STM experiments and long-term STML measurements in particular. Moreover, internal degrees of freedom of molecules may have low energy barriers and can be excited at room temperature [196] or even at liquid nitrogen temperatures [197]. In my diploma studies, I have observed thermally driven switching of single molecule junctions in STM at temperatures as low as 5.3 K [198, 199]. So in order to achieve well defined experimental conditions over extended periods of time, low temperatures are inevitable when handling single molecules. Welcome side-effects are the significantly higher energy resolution for spectroscopy and very low thermal drift of the entire microscope.

In this setup, a commercial Joule-Thomson style cryostat is used, which reaches a base temperature of 1.3 K as measured at the STM. The cryostat design was conceived in our group by Zhang *et al.* [103] and is now manufactured and sold by *CryoVac*. It is optimised for long standing times between refilling the coolants, facilitating extended uninterrupted measurements and saving especially on the expensive liquid helium.

3.3.1 Basic structure

The cryostat consists of three stages that are shown in Fig. 3.15: the liquid nitrogen (LN_2), the liquid helium (LHe) and finally the Joule-Thomson (JT) stage. The first two stages are bath cryostats simply holding cryogenic fluids and deploying cooling power by evaporating the coolant. They are concentrically nested because nitrogen features a much higher evaporation enthalpy than helium [200, 201] and is also much cheaper. It is, therefore, used to shield the liquid helium stage from the powerful ambient heat radiation.

The JT stage utilises the Joule-Thomson effect, which causes non-ideal gasses to cool upon adiabatic expansion below a critical temperature. Helium gas is injected into the JT inlet at room temperature and an absolute pressure of 3 to 7 bar. It is then pre-cooled by counter-flow heat exchangers and the liquid helium bath cryostat. Finally, it is expanded through a thin capillary and collected in a small pot (the JT pot), which is at a pressure of around 1×10^{-2} mbar. To maintain the pressure difference across the capillary, a pump is required that constantly removes helium from the JT pot, which also gives rise to evaporation chill. For that purpose, a rotary vane pump with a throughput of $20\,\mathrm{m^3\,h^{-1}}$ is attached to the low-pressure exhaust. The connection to the pump is a 3 m long PVC tube that is at one point embedded in a block of concrete standing on the floor to decouple the cryostat against the vibration of the pump.

A major design goal is minimising the consumption of the cryogenic fluids. Heat conduction through convection does not occur in UHV, and heat conduction through the material is minimised because the only connection to the helium tank is a thin walled stainless steel tube. The only heat source left is radiation, which is kept away from both tanks with additional heat shields that are referred to as radiation shield bells and plates in Fig. 3.15. These comprise additional reflective surfaces in the path of the heat radiation and are cooled by the evaporating nitrogen and helium gas, respectively. This effective technique utilises the cooling power of the evaporating gas instead of consuming any cryogenic fluid for cooling the radiation shields. To further reduce the liquid helium consumption, all accessible surfaces of the radiation shields as well as the helium tank were additionally covered with highly reflective aluminium foil.

The JT stage is designed to reach 600 mK if operated with a turbomolecular pump (TMP) and ^{3}He as a working gas. To reach such low temperatures, heat transfer from the liquid helium stage to the JT stage has to be prevented, so that these two are well isolated from each other by design. In this setup, the minimum reachable base temperature is considerable higher due to the use of a rotary vane pump and ^{4}He. Therefore, this thermal isolation does not need to be as proper and was reduced, because it imposed a practical problem. Proper thermal decoupling of these stages results in long cool-down times during system cool-downs and after inserting new samples

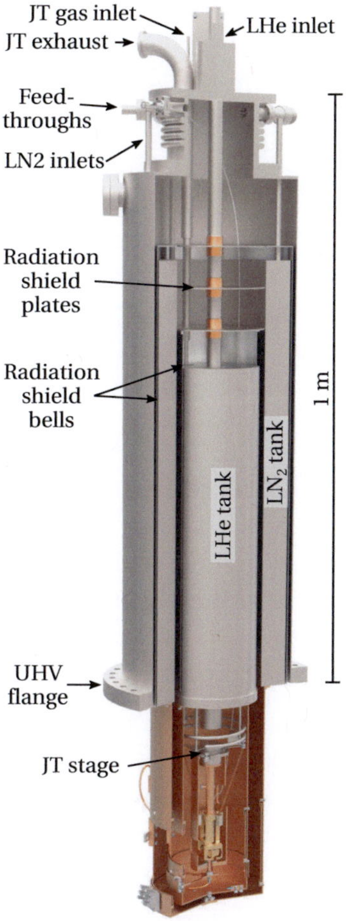

Figure 3.15: Overview of the cryostats structure featuring the feedthroughs and inlets at the top, the three cryogenic stages and the experimental section at the bottom.

because the Joule-Thomson effect only yields substantial cooling power at already low temperatures, and the cooling power of the liquid helium stage is required to reach this regime first. To remedy this issue, the JT and liquid helium stage were thermally connected via a 6 mm thick and 25 mm long threaded alumina cylinder. This material features a high thermal conductivity at high temperatures, which still amounts to a few $\mathrm{W\,m^{-1}\,K^{-1}}$ down to 10 K, but then steeply drops to $10^{-2}\,\mathrm{W\,m^{-1}\,K^{-1}}$ at 1 K [202, 203]. Therefore, the alumina block acts as a thermal switch, which connects the two stages to enable fast cool-downs at elevated temperatures, and then isolates them as soon as the temperature is low enough for the JT stage to run on its own. It was found that installing this thermal connection almost halved the system cool-down time but did not result in a measurable increase of the base temperature.

In the future, it might be of interest to perform variable temperature experiments. For that, a small 1 kΩ resistor in a surface mount device package was placed at the JT stage, that can act as a heater. The idea is to drive this resistor with a spare output of the *Nanonis* electronics because that way, a feedback loop can be implemented which adjusts the heater power to maintain a certain temperature. However, the stability of such an approach has not been tested yet, as no temperature variation was necessary for the measurements presented in this work.

3.3.2 Radiation shields and parking mechanism

The experimental section at the bottom of the cryostat accommodates the STM, which needs to be suspended and shielded from heat radiation. There also needs to be a mechanism that can be operated *in-situ* and mechanically fixes the STM during tip and sample transfers, which is referred to as *parking mechanism.*

As shown in Fig. 3.16, three nested radiation shields are installed at the bottom of the cryostat, each of which is bolted to the liquid nitrogen, liquid helium and JT stage, respectively. Their purpose is to reflect as much of the incoming heat radiation as possible, and the rest, which is inevitably absorbed, is to be efficiently conducted to the respective cryostat stage instead of warming up the shield. Therefore, they were chosen to be fabricated from 2 mm thick OFHC copper for its excellent heat conductance. After fabrication, they were ground, polished to a mirror finish and chemically coated with silver for maximum reflectivity.

It is possible to access the STM for tip and sample transfers by opening sliding doors in the outer two radiation shields. Each of the doors consists of a concentric cylinder within the shield with a large opening that can be brought in line with a corresponding opening in the shield by turning the inner cylinder with a wobble-stick. The inner

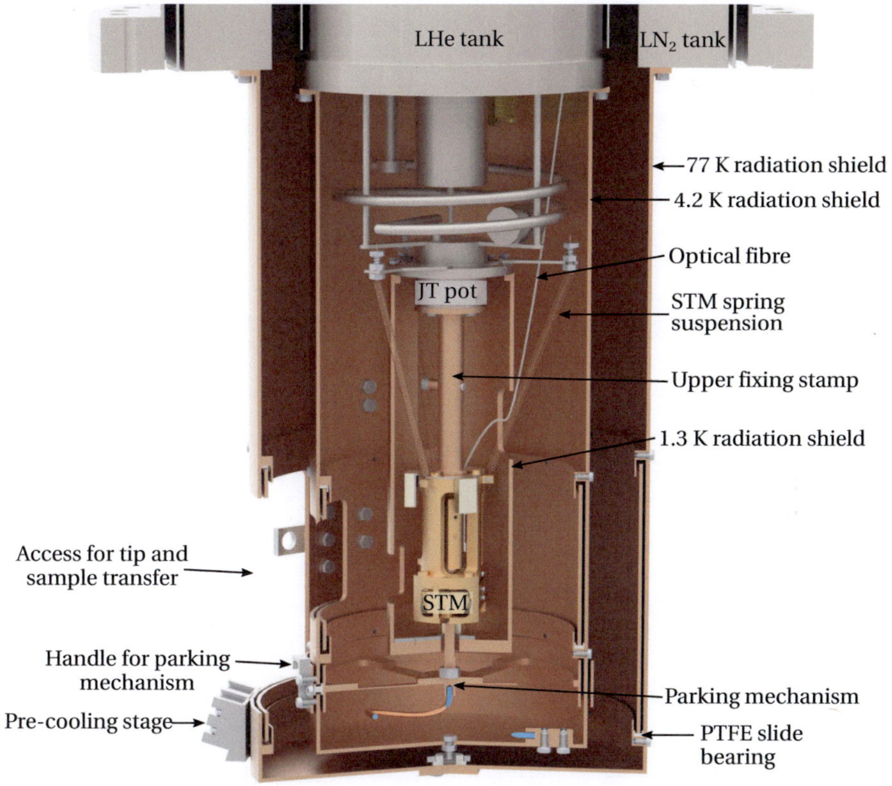

Figure 3.16: Design of the bottom section of the cryostat accommodating the STM. The sliding doors are depicted in their open configuration.

cylinders are supported by polytetrafluoroethylene (PTFE) slide bearings at the top and bottom, which exhibit low friction even in UHV, so that they do not get stuck. Special care was taken to ensure that the doors are completely light-tight when closed. The inevitable gap between the inner cylinder and the shield was sealed with a circumferential L-shaped cover on the inside, which was coated with a highly absorbing black carbon paste. That way, heat radiation cannot pass through the gap by multiple reflections anymore. The inner cylinders were thermally coupled to the shield with thick copper braids to prevent them from heating up. The braids are brazed into copper terminals on both ends, which are screwed to the shield and the door respectively and additionally serve as a handle for the wobble-stick.

As the emitted power of heat radiation scales with T^4 according to the Stefan-Boltzmann law, the heat radiation from the 4.2 K helium shield is negligible compared to the heat radiation emitted from the ambient or the 77 K nitrogen shield. So the innermost

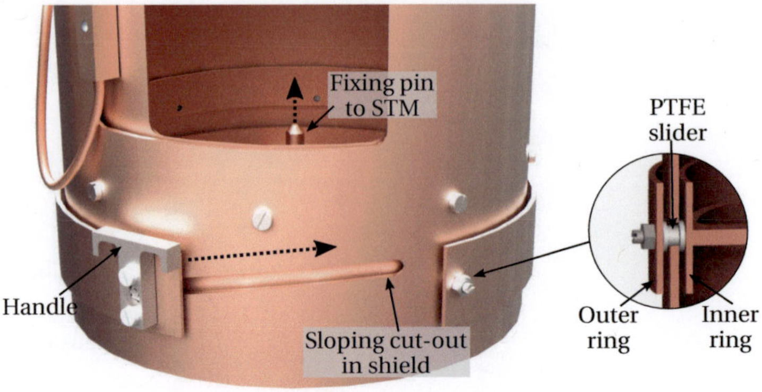

Figure 3.17: Parking mechanism used to fix the STM during tip and sample transfers. A section of the front radiation cover was cut in the drawing to expose one of the five cut-outs by which the assembly is guided. Moving the handle to the side causes the fixing pin to move upwards against the STM. The inset exposes the PTFE bearing within the cut-out.

shield is exposed to almost no radiation anymore and mainly exists for the eddy current brake of the STM. Therefore, this shield was not equipped with a sliding door.

The parking mechanism, which is used to mechanically fix the STM if necessary, can be seen in Fig. 3.16 below the STM and in detail in Fig. 3.17. It consists of a concentric ring within the 4.2 K radiation shield below the door, that supports a central pin, which moves upwards once the mechanism is engaged. The pin then pushes the STM from below against the upper fixing stamp. In this, there are two cone-shaped pins which are located off-centre and engage corresponding holes in the top of the STM. Therefore, both translation and rotation of the STM are prevented. The lifting action of the bottom pin is achieved by rotating the ring from outside the 4.2 K helium shield. Along its perimeter, the supporting ring inside the helium shield is connected to another ring outside the shield through five fasteners poking through elongated, sloped cut-outs in the helium shield. One of those cut-outs is exposed in Fig. 3.17 by removing a section of the outer ring. The inner and outer rings are extruded and, thus, cover the cut-outs to prevent heat radiation from entering the shield. The connecting fasteners are enclosed by a set of PTFE slide bearings, shown in the inset of Fig. 3.17, and can slide within those cut-outs. Once the handle is turned to the side, the assembly of outer ring, inner ring and fixing pin simultaneously perform a rotation and a slight upwards movement. The fixing pin is threaded at its bottom and screwed into the inner supporting ring. Therefore, its height can be adjusted such that, once lifted, it holds the STM firmly in place without requiring too much force.

In the parked state, the central pin also provides thermal coupling to the liquid helium stage to further speed up the cool-down after sample transfer. For measurements, the STM is released from its parking position and then suspended by three springs. Because the thermal connection of the STM to the cryostat through the springs is insufficient in the suspended state, three short 125 μm silver wires establish a thermal contact between the STM and the JT stage.

3.3.3 Wiring

Heat transfer from ambient into the STM through the electrical wiring is prevented by two measures. First, all cables except the one carrying the tunnelling current were interrupted at both the liquid helium and JT stage and plugged into pins that are well coupled to the respective cryostat stages. The cable for the tunnelling current was not interrupted for better noise immunity, and instead pressed onto the cryostat bottom with a screw. Second, thin stainless steel cables were used, which provide decent thermal insulation. The scanner piezo drive signals, tunnelling current and voltage as well as the additional sample contacts and thermometer leads were wired with shielded twisted-pair cables (*Cooner Wire* AS631-2SS) down to a connector terminal at the JT stage. The total resistance of each of these lines is around 270 Ω at room temperature. The coarse motion, however, requires high peak currents for the quick slip motion. Therefore, these lines were wired with thicker cables (AS636) that have a resistance around 40 Ω per line.

From the connector terminal at the JT stage to the STM, the coarse motion signals and the tunnelling voltage are carried by polyimide insulated copper wires. The wire for the tunnelling voltage is the main thermal connection of the sample receptacle within the STM. To achieve optimal thermal coupling to the JT stage despite the electrical insulation, this wire was wound approximately 100 times around a small copper block and glued into place. The copper block was then screwed to the JT stage. That way, the contact area of the wire is high enough to quickly conduct the heat of newly inserted samples through the electrical insulation to the cryostat, and the thermalisation of the sample receptacle does not need to rely on the well insulating piezo actuators.

3.3.4 Performance

The base temperature of the cryostat, as measured at the back of the STM body, is approximately 1.3 K. When the JT stage is not running, a temperature of 4.4 K is reached, which is very close to the temperature of the liquid helium. This indicates that only very little heat radiation is introduced into the system. Representative cooldown times are shown in Fig. 3.18. When the system has been serviced and needs to

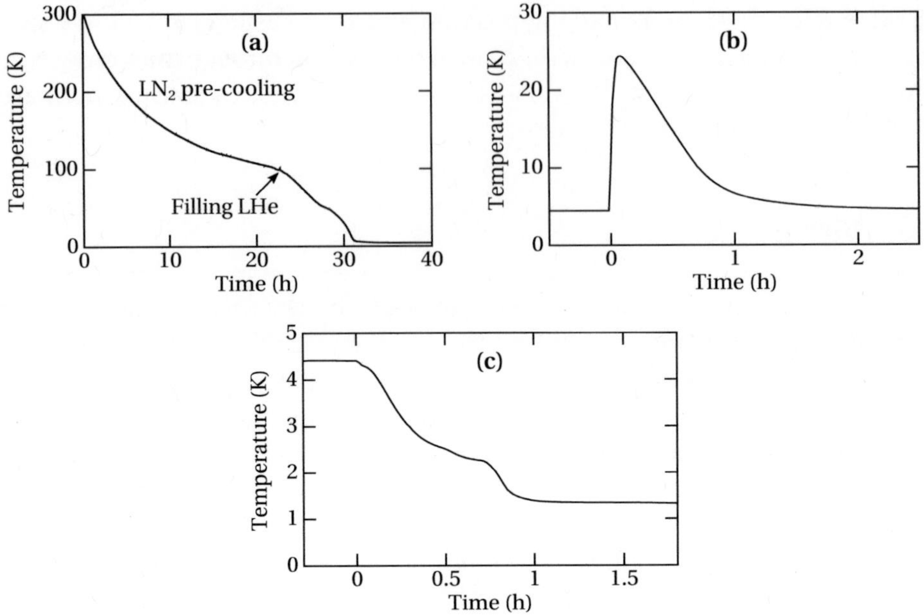

Figure 3.18: Representative cool-down curves of the finished setup as measured at the STM body (a) from room temperature to 4.4 K after maintenance, (b) after transferring a new sample into the STM, (c) from 4.4 to 1.3 K.

be cooled down from room temperature, the helium bath cryostat is first pre-cooled with liquid nitrogen until the STM is at 100 K, which takes one day. Subsequently, the liquid nitrogen is removed and liquid helium is filled, which cools the STM to 4.4 K within another 12 h. Then, when the STM is already cool and a new sample is inserted, the temperature first rises to about 25 K and then drops back to the initial temperature within 2 h, see Fig. 3.18(b). This enables a high sample throughput and short waiting times. Finally, when the JT stage is activated, Fig. 3.18(c) demonstrates that it takes about 1 h to reach the base temperature.

Refilling the cryogenic fluids interrupts any STM measurement due to unavoidable mechanical disturbances. Therefore, it is highly beneficial that the liquid nitrogen and liquid helium tanks only need to be refilled every 4.5 and 7.5 days, respectively.

3.4 UHV chamber

The preparation and investigation of clean samples require UHV, which conveniently also provides isolation for the cryostat with regard to convection. For that reason, the

STM and all the sample preparation equipment is installed in a UHV chamber. The whole setup has a comparably small 1.5 m × 2.4 m footprint and a height of 2.3 m. The design was adapted from Balashov *et al.* [204]. It easily fits into a regular-sized laboratory and is equipped with all necessary tools for preparing samples *in-situ* and transferring them into the cool STM. The UHV system is made up of three independent chambers that are referred to as STM, preparation and load-lock chamber as indicated in Fig. 3.19. Of those, only the load-lock chamber is frequently vented to transfer samples or tips from ambient conditions into the system, and the vacuum in the other chambers is sustained at all times unless maintenance is required. The chambers are mounted on a frame of aluminium extrusions, which is suspended by pneumatic vibration isolators made by *Newport*. These decouple the machine from vibrations of the floor. Apart from that, the setup does not rely on further heavy or active dampening stages, rendering costly and time-consuming structural alteration of the laboratory unnecessary.

3.4.1 Vacuum components

Each of the three chambers is equipped with their own turbomolecular pump (TMP) and can be isolated from each other by means of gate valves. Venting, maintaining and pumping down again any of these chambers is therefore possible without compromising the vacuum in any of the other ones. The UHV chambers themselves were manufactured from stainless steel by *VACOM* and assembled following usual guidelines to prevent contamination of the inner surfaces with anything that exhibits long-term outgassing. Any custom components that reside inside the UHV have been build exclusively from materials that have a very low vapour pressure ideally up to the bake-out temperature. This prohibits the use of rubber, most plastics except polyimide, PTFE and PEEK, as well as many soft metals like tin, zinc, lead and indium including their alloys.

The UHV in each chamber is established by the aforementioned TMPs. After a bake-out procedure, for which the preparation chamber is heated to around 200 °C and the STM chamber to 100 °C for a few days while the TMPs are running, the base pressure in these chambers is around 1×10^{-10} mbar. During STM operation, however, the vibrations caused by the high rotation speed of the TMP rotors would disturb the measurement. Therefore, all TMPs have to be switched off during measurements. The UHV in the STM and preparation chambers is then maintained by mechanically quiet ion getter pumps. The preparation chamber is additionally equipped with a titanium sublimation pump, which is operated periodically.

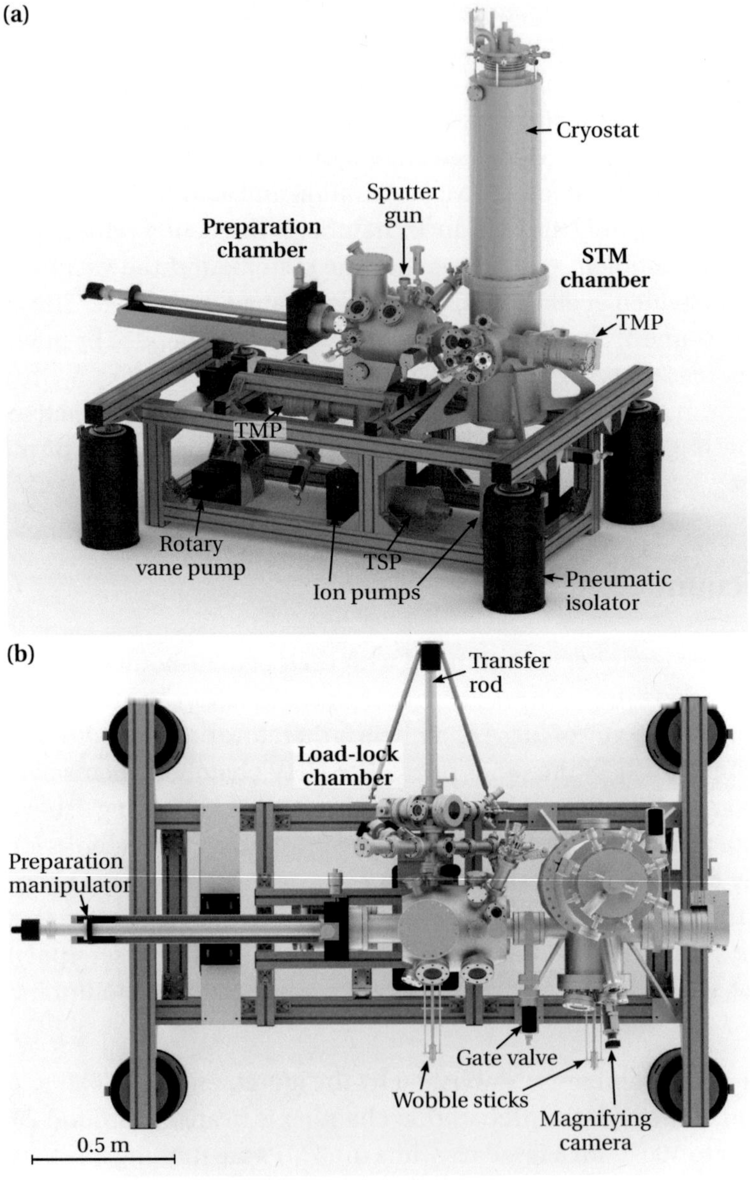

Figure 3.19: CAD rendering of the UHV setup with its three independent chambers (naming in bold letters). (a) Front view highlighting mostly vacuum components and (b) top view indicating mostly handling components. TSP stands for titanium sublimation pump.

The TMPs require a rough vacuum at their outlets, which is provided by a separate vacuum line. Each TMP can be separated from this rough vacuum line by closing a diaphragm valve. This is necessary to prevent oil vapour from diffusing into the pump when it is switched off, or if another previously vented chamber needs to be pumped and the pressure in the rough vacuum line rises temporarily. The pressure inside the rough vacuum line is kept around 1×10^{-2} mbar by a single rotary vane pump. The TMPs deliver only a small amount of gas into this line, so that the rotary vane pump does not need to run all the time but rather with a duty cycle of around 20 %. A custom circuit with a relay switches the pump on and off if the pressure in the rough vacuum line rises above or drops below user defined setpoints, respectively. The line is buffered with a 20 l vacuum barrel to increase the switching times. Without additional dampening, the vibration of the rotary vane pump is notably transferred onto the chamber. To protect the sensitive TMPs from that vibration, the feed line to the rotary vane pump was moulded into a 20 kg block of concrete standing on the laboratory floor.

The residual pressure in the UHV chambers is monitored with hot ion gauges in the preparation and STM chambers, and with a cold ion gauge in the load-lock chamber, respectively. The latter is used because the pressure in the load-lock chamber often exceeds the range in which hot ion gauges operate, and cold ion gauges are much more tolerant towards accidental venting. To locate leaks and assess the composition of the residual gas in the chambers, the preparation chamber is also equipped with a quadrupole mass spectrometer.

3.4.2 Handling and sample preparation

Samples and tips are mounted on carrier plates, similar to the standardised *Omicron* plates, that feature a knob at which they can be grabbed, and are moved through the machine by three different kinds of manipulators, which are operated from outside the chambers. The position of those manipulators is indicated in Fig. 3.19(b).

New samples or tips are usually inserted into the chamber through the quick access door in the load-lock chamber. There, they are placed into the head stage of a magnetically coupled transfer rod which accommodates up to three carrier plates, see the magnified detail in Fig. 3.20. These plates can be pushed into the preparation chamber where they are transferred onto the sample stage of the preparation manipulator or into a storage stage. This is done with another magnetically coupled manipulator referred to as wobble-stick, which is equipped with a mechanism to securely grab and release the knobs of the carrier plates and provides four degrees of freedom. The preparation manipulator is sealed with a flexible metal bellow, and provides precision

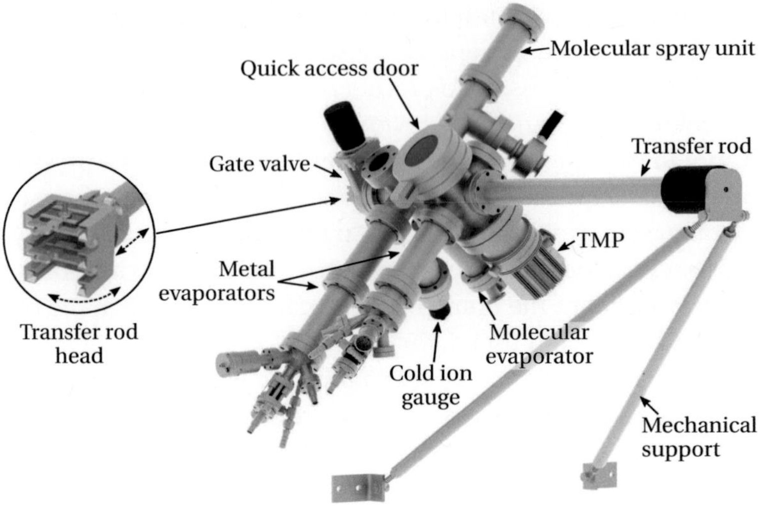

Figure 3.20: Design of the load-lock chamber, which is attached to the preparation chamber via a gate valve. It is used for molecular deposition, metallisation of the mirror tips, and to introduce samples and tips into the UHV.

positioning and a long travel reaching into the STM chamber, which is used to transfer carrier plates there. When extended into the STM chamber, another wobble-stick is then used to transfer the plates either into a pre-cooling stage connected to the liquid nitrogen radiation shield, or into the STM.

The preparation manipulator is supposed to be advanced along its long direction by manually turning a crank at its end. This is not only inconvenient but puts the long and by design poorly supported manipulator under high stress, as it is impossible to rotate the crank without applying sidewards force. The sample stage inside the vacuum is easily shaken that way, potentially leading to loosing the sample or scratching the walls at the bottleneck between the chambers. This manipulator was therefore fitted with a stepper motor, which is driven by a home-made *Arduino* based controller, and advances the sample stage smoothly through the chamber.

The sample preparation occurs in the preparation and the load-lock chamber. For that, the sample stage of the preparation manipulator is heatable and coolable, and can be placed in front of a sputter gun for cleaning or a metal evaporator for depositing thin films. The load-lock chamber, which is shown in detail in Fig. 3.20, was equipped with two metal evaporators loaded with silver and chromium, respectively, to metallise the mirror tips. This chamber was chosen for this purpose because only the transfer rod features the necessary rotational degree of freedom to coat the mirror tips from both sides with no obstruction below or above. Two molecular deposition devices were also integrated into this chamber because these comparably dirty techniques might otherwise compromise the UHV of the preparation chamber. Molecules which can be sublimed are deposited onto the sample from a home-made molecular evaporator developed by Tobias Engelhard in the scope of his masters thesis [205]. Many complex organic molecules tend to decompose prior to sublimation, though. These can be deposited using an inexpensive and simple molecule spray unit, which was developed in our group and described in detail by the master theses of Timo Frauhammer and Loïc Mougel [206, 207]. Both of these devices have already been used to deposit large organic and metal-organic molecules onto clean Au(111) surfaces and corresponding STML measurements are pending.

4 Performance tests on Au(111)

The newly created experiment with all its components is fully operational and different aspects of its performance are demonstrated in this chapter. First, the usability of the mirror tips for STM is showcased on a well-known Au(111) surface and the — both short-term and long-term — stability of the STM is characterised. So far, the γ-STM was tested with various mirror tips, which were treated with careful voltage pulses and controlled surface contacts, and also metallic tips that were conventionally fabricated by electrochemically etching a tungsten wire. High quality STM images with low drift have been obtained with both types of tips. Then, electroluminescence measurements on clean Au(111) are presented, which reproduce different effects known from the literature in great detail. The mirror tips do almost always yield an intense light signal. The spectral composition of the emitted light revealed that sharp tips with highly confined plasmon modes can be achieved with these coated polymer tips. So far, this aspect of the performance varied greatly among different mirror tips, though.

A Au(111) single crystal purchased from *MaTecK* was chosen for the performance tests, because its surface is easy to prepare and well-known. The gold substrate also exhibits a strong plasmonic response and is a widely studied system for STML. The surface was prepared *in-situ* by repeated Ar^+ ion sputtering and annealing to 550 °C.

4.1 STM performance

4.1.1 Image quality and resolution

The imaging capabilities of a mirror tip are demonstrated in Fig. 4.1(a). On the overview scan, the atomic step edges are sharply reproduced (see profile in (b)) and the distinctive herringbone surface reconstruction [208] is clearly visible. The latter is clarified by applying a different contrast to the inset area. Overall, the image is of the same quality as could be expected from a conventional STM tip. One distinct periodic disturbance with an amplitude of 2.5 pm, which was externally coupled into the STM at the

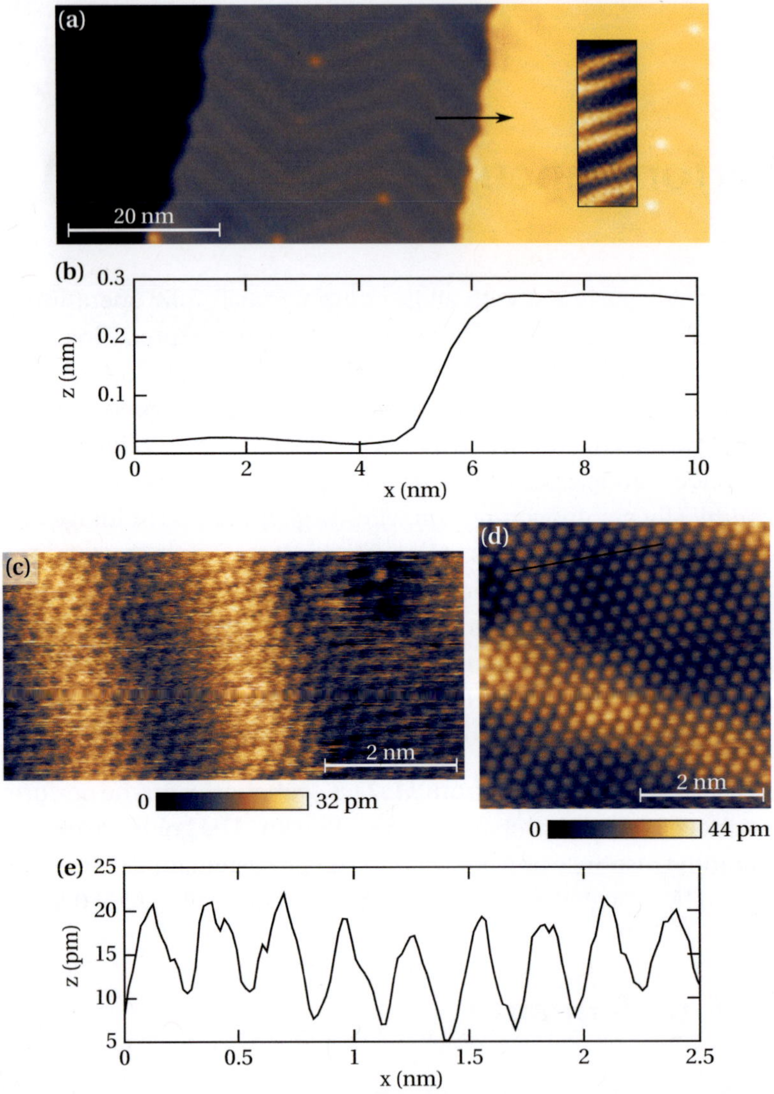

Figure 4.1: STM topographs of a Au(111) surface. (a) Atomic terraces recorded with a mirror tip in constant current mode at 1.4 K and tunnelling parameters of 2.5 V and 19 pA. The inset highlights the herringbone reconstruction with adapted contrast. Some Fourier-filtering was applied (see text). (b) Height profile across the atomic step edge along the arrow in (a). (c) Topograph with atomic resolution taken at 4.4 K with a mirror tip at tunnelling parameters of 1 V and 54 pA, and (d) using a tungsten tip at 150 mV and 500 pA. Both images are unfiltered. (e) Profile along the black bar in (d).

time of this measurement, was removed from the picture in Fig. 4.1(a) using narrow-band Fourier-filtering, as it is not relevant for assessing the performance of the mirror tip. In Fig. 4.1(c), atomic resolution was achieved with another mirror tip, reproducing the well-known hexagonal symmetry of the (111) surface of face-centred cubic crystals. This excellently corroborates the usability of the DLW tip for STM, despite the larger macroscopic tip radius compared to electrochemically etched metal tips.

One issue, which will be discussed later, is the limited workability of the mirror tips, though. Using a conventionally prepared tungsten tip, which can be worked more intensely until it features the desired imaging capabilities and stability, the image in Fig. 4.1(d) was recorded. It features higher resolution and no artefacts from changes of the tip configuration, and the profile presented in Fig. 4.1(e) demonstrates the good stability of the γ-STM.

The relation between applied voltage and displacement of the scanner piezo tube at low temperatures is, at first, unknown and needs to be calibrated. For the lateral movement, the observed inter-atomic distances along the densely packed directions in Fig. 4.1(e) were compared to the values that are known from x-ray diffraction measurements. The lattice constant of gold is $a = 4.07\,\text{Å}$ [209], and the nearest-neighbour distance on (111) facets is $d_{\text{NN}} = a/\sqrt{2} = 2.87\,\text{Å}$. Thus, the calibration constant in the *Nanonis* user interface was set such that this periodicity is reproduced on the recorded images. A similar procedure was followed to calibrate the vertical movement of the scanner tube, i.e., by comparing the measured step height in Fig. 4.1(b) with the distance of individual (111) layers, which is $d_{\text{terrace}} = a/\sqrt{3} = 2.35\,\text{Å}$.

4.1.2 Short-term stability

The mechanical stability of the tunnelling junction was quantified by the following measurement. A mirror tip was approached into tunnelling contact at 4.4 K and stabilised at a sample bias voltage of 1 V and a tunnelling current of 50 pA. The feedback loop then constantly corrects for changes of the tip-sample distance due to mechanical noise entering the STM. After the piezo creep has decayed, the time trace of these position corrections therefore resembles the vibration of the tip against the sample within the accuracy of the feedback loop. Such a time trace was decomposed into its frequency components by a Fourier transform, which is presented in Fig. 4.2. The vibration spectrum exhibits the well known $1/f$ characteristic from fluctuations of the piezo drive voltage, and a few peaks that stem from specific mechanical oscillation modes of the cryostat and the frame. The integrated vibration noise amounts to a root-mean-square (RMS) of 1.5 pm, which can be considered decent compared to other published

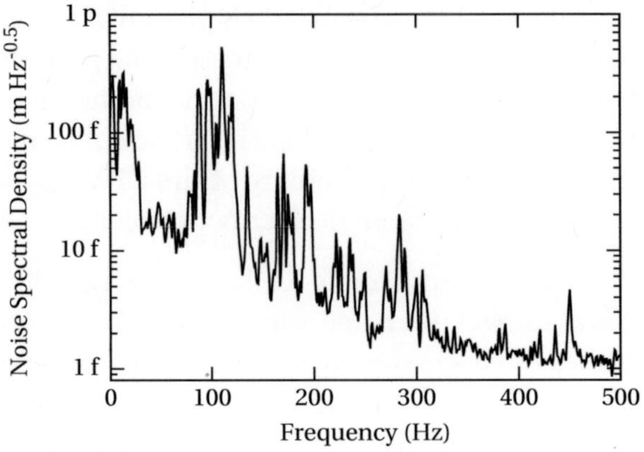

Figure 4.2: Vibration noise spectrum recorded with a mirror tip on Au(111) in usual measurement conditions with activated feedback loop at 4.4 K.

RMS noise figures of 0.2 to 1 pm [101, 103, 104, 204, 210, 211], taking into account the wide coarse motion range of the γ-STM — which inevitably makes the design less compact than others — and the lack of any heavy or active dampening stages.

4.1.3 Long-term stability

Electroluminescence measurements with systematic variation of the tip position or the tunnelling parameters take a long time, and it is critical that the tip rests at precisely the indented location during this time to avoid unwanted variation of the experimental conditions or excessive distortion on spatially resolved maps. One source of undesired tip movement is thermal drift caused by minute variation of the temperature and, thus, thermal expansion of the different components within the STM. Additionally, the output voltage of the controller electronics may change over time due to changes of the temperature in the laboratory, which causes an apparent position shift as well. In order to characterise the typical long-term stability of the γ-STM, a tungsten tip was approached on a Au(111) sample at 1.35 K and the piezo creep was allowed to settle for several hours. Then, a series of images was acquired at the same scanning frame over the course of 15 h, some of which are presented in Fig. 4.3(a). They show a flat section of the Au(111) surface with a distinctive contaminant, which apparently moves over time because the tip and sample slowly move with respect to each other. The precise location of the contaminant within each picture was determined with a 2D cross-correlation algorithm and reveals the drift trajectory in Fig. 4.3(b). By averaging over the apparent position shifts between each subsequent picture, a mean drift velocity of merely 52 pm h^{-1} was found. As apparent from Fig. 4.3(c), this coincides

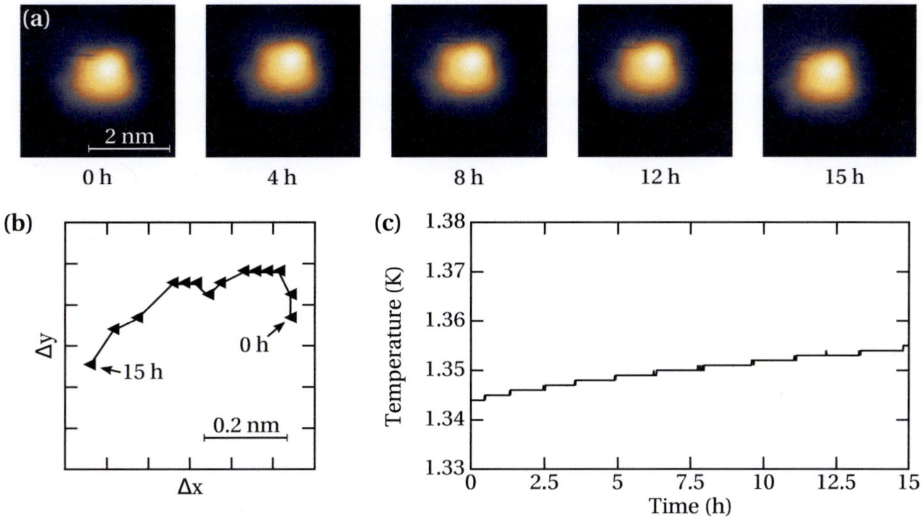

Figure 4.3: Apparent shift of the lateral position of a surface contaminant over time due to minute thermal drift. (a) Image series taken at the same frame position over the course of 15 hours at 300 mV and 3 pA. (b) Relative position shift extracted from the images, which were taken every hour. (c) Time-evolution of the temperature at the STM during the measurement.

with a slow increase of the temperature at the STM of 0.73 $\mathrm{mK\,h^{-1}}$. All of this can be considered very stable. For other STML setups, drift velocities from 100 to 200 $\mathrm{pm\,h^{-1}}$ [82, 87, 212] or even up to 3 $\mathrm{nm\,h^{-1}}$ [132] have been reported so far, and most authors do not even publish this figure. The superior stability of the γ-STM compared to other STML setups is probably to be attributed to the radiation shields, which are completely light tight during the measurement, rendering the temperature of the STM extremely stable.

During the measurement in Fig. 4.3, also an average vertical drift of 11 $\mathrm{pm\,h^{-1}}$ was observed. However, this figure is not as important, because the feedback loop is usually never switched off that long and, therefore, constantly corrects for the vertical drift.

The inherent long-term stability of the γ-STM is excellent. However, the mirror tips, which have been employed so far, often exhibited problems in that regard. First of all, silver is a quite soft material in itself. Second, the vapour-deposited silver layer is in a disordered, meta-stable configuration with numerous grain boundaries and dislocations because it is impossible to anneal the coated mirror tips without affecting the underlying polymer. Therefore, atomic rearrangements at the tip happen frequently when a mirror tip is in tunnelling contact, especially at elevated tunnelling currents or bias voltages, and it is tedious to treat the tip until it is sufficiently stable, especially for long measurements. This issue is amplified by another problem, that is the already mentioned limited workability of the mirror tips. It has indeed proved possible to shape the mirror tips by carefully applying voltage pulses or by controlled sample contacts. However, this can only be performed a limited number of times before the coating wears off and has to be reapplied.

4.2 Light collection

4.2.1 Single luminescence spectra

Fig. 4.4(a) shows light emission spectra obtained with three different mirror tips from a clean Au(111) surface at the same bias voltage, similar current setpoints and, thus, similar tip-sample distances. These spectra are similar to luminescence spectra from the literature, which were acquired with metal tips. Depending on the tip shape, vastly different intensity distributions with few or many broad peaks across almost the entire visible wavelength range and near-infrared are produced. The black curve in Fig. 4.4(a) is made up of at most three peaks, whereas the red curve features such a dense spectrum that barely any individual peaks can be distinguished. This can be attributed to a very blunt tip, which produces a smaller mode spacing according to Eq. 2.14. Also, blunt tips are likely to be asymmetric or feature other micro-tips close to the tunnelling junction, which causes several plasmon resonances with different lateral confinements [114]. In contrast, sharper tips feature fewer plasmonic peaks, which occur already at lower wavelengths.

Comparison of the intensities measured with different mirror tips is difficult, as they always pertain to both the emission intensity of the tip and the collection efficiency of the mirror, and the latter might vary among different mirror tips. In the literature, it is stated that for sharp tips, the intensity of individual resonances is expected to be higher compared to blunt tips. This does not necessarily need to apply to the integrated intensity, as no theoretical work so far considered the numerous resonances of a blunt tip, which may add up to a brighter total intensity. This statement seems to be compatible with the spectra presented in Fig. 4.4(a), assuming that the respective collection efficiencies were similar. However, as the intensity maximum of sharper tips tends to be at lower wavelength, their spectrum also goes better with the response curve of the detection setup. Fig. 4.4(b) shows the same spectra but with the detector response taken into account, and it is apparent that the main peak of the red curve is almost as high as the peak of the black curve. Analysing the spectrum of sharper tips is still easier as the detection rates are higher and the spectra are more well-defined.

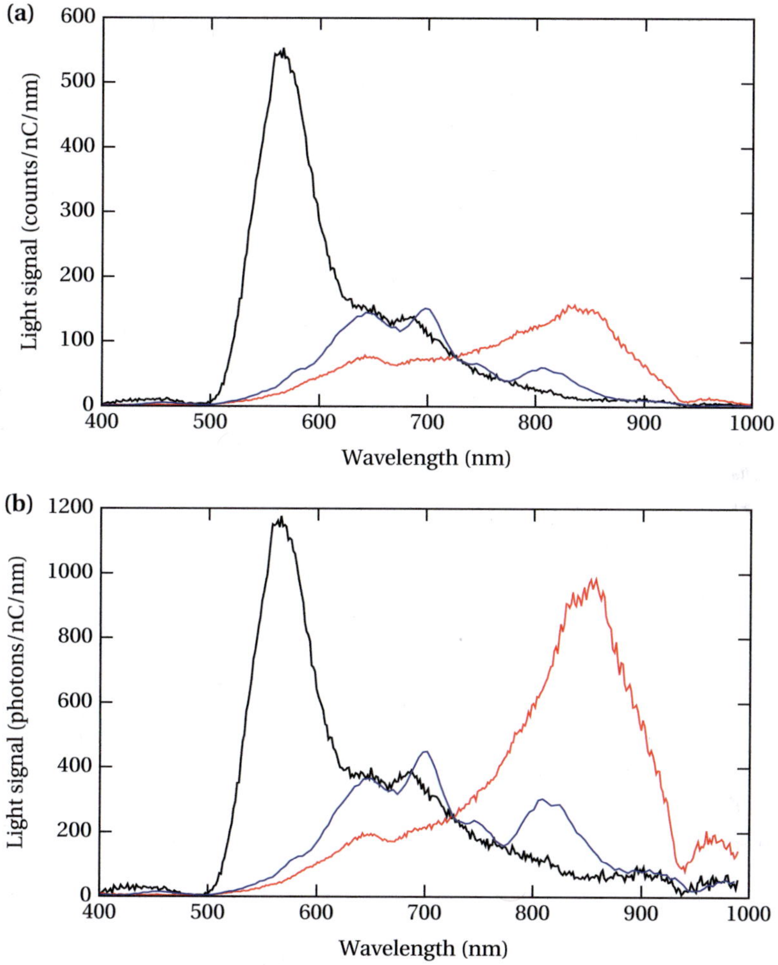

Figure 4.4: Electroluminescence spectra recorded on a Au(111) surface with different mirror tips at a sample bias of 3.5 V each. Tunnelling currents were 1 nA (black and blue) and 5 nA (red). The spectra were normalised with respect to the tunnelling current, the exposure time and the spectroscopic bin width. (a) The spectra as measured by the CCD camera, (b) the same spectra as in (a), but corrected for the detection response curve shown in Fig. 3.4.

Sharp tips also offer a higher spatial resolution, as the gap plasmon is more confined to the tunnelling junction.

Another reason for the high intensity of the the black curve's main resonance in Fig. 4.4, apart from sharpness considerations, can be found in the fact that it roughly coincides with a minimum of the imaginary part of the dielectric function of the gold substrate, which describes the dielectric material losses. At lower wavelengths, plasmon resonances are attenuated by an interband transition, and at higher wavelengths emerging intraband transitions cause the same effect [213–215]. If a resonance around 2 eV is excited, however, the energy of the mode is less likely to be dissipated and, therefore, more likely to couple into the electromagnetic far-field. Also a more rigorous theoretical work predicted a resonance effect at 2.1 eV for gold substrates [46]. From the black curve in Fig. 4.4, a resonance in this suppressed wavelength range around 450 nm is actually apparent, but it is strongly attenuated despite the sufficiently high excitation energy provided by the tunnelling bias of 3.5 eV.

Of the mirror tips employed so far, a few featured a sufficient sharpness to produce a plasmon spectrum with only a few, bright resonances. Whenever this was not the case, it was tried to sharpen the tips *in-situ*. This mainly improved the stability and STM resolution and also affected the total emission intensity, but never resulted in spectra with fewer peaks. This is because improvements of the STM performance can achieved by changing the configuration of the topmost atom. But in order to substantially change the properties of the plasmonic cavity, modification of the tip in the order of several nanometres around the tunnelling junction is required, which is more difficult to achieve by voltage pulses and sample contacts without tearing off the entire metal coating. If a mirror tip did not feature the desired emission properties, it proved possible to restore or improve it by coating it with another layer of silver.

4.2.2 Voltage maps

The light intensity at the detector is high enough to easily facilitate automatic, systematic variation of a tunnelling parameter and to investigate its influence on the light emission in a reasonable time. On a clean Au(111) surface, such an experiment was performed to characterise the voltage dependence of the light emission at constant current. In both maps shown in Fig. 4.5, each line represents an optical spectrum recorded at a different sample bias, while the feedback loop was activated. The light intensity as detected by the CCD at each wavelength is colour coded. The blue curve in Fig. 4.4 is a spectrum taken from Fig. 4.5(a) at a bias voltage of 3.5 V, and the black curve in Fig. 4.4 is taken from Fig. 4.5(b). In both maps, three regimes can be clearly distinguished in the probed voltage range. At constant wavelength, light emission can only be detected above a certain voltage threshold for both polarities. This threshold is indicated by the grey, dotted lines in Fig. 4.5 and corresponds to the so-called quantum cut-off [28]. At these tunnelling parameters, the gap plasmons are predominantly excited by inelastic tunnelling of single electrons, which only occurs if the energy of the tunnelling electrons eV is above the excitation energy. At intermediate voltages, the detected light intensity is highest, and above a certain absolute value, the intensity quickly decreases. As apparent from those two maps obtained with different tips, the upper voltage boundary depends both on the polarity and the tip.

In the literature [117, 128], this has been attributed to the change of the tip-sample separation during such an experiment, because the feedback loop continuously adjusts the vertical tip position in order to achieve a constant tunnelling current. During the measurement of the voltage maps, the vertical tip position was recorded for every spectrum. However, minor tip changes and significant piezo creep during these measurements superseded the distance variation caused by the bias variation, and rendered it impossible to extract the undisturbed changes of the tip-sample separation during these particular experiments. Therefore, the voltage-distance characteristic at constant current was later measured again with another tip. In the measurement presented in Fig. 4.6(a), the feedback loop was set to 50 pA and the bias voltage was varied between 12 mV and 6 V for both polarities. At low bias voltages, the tip was close to the sample in order to achieve the set tunnelling current, and increasing the voltage caused the feedback loop to lift the tip.

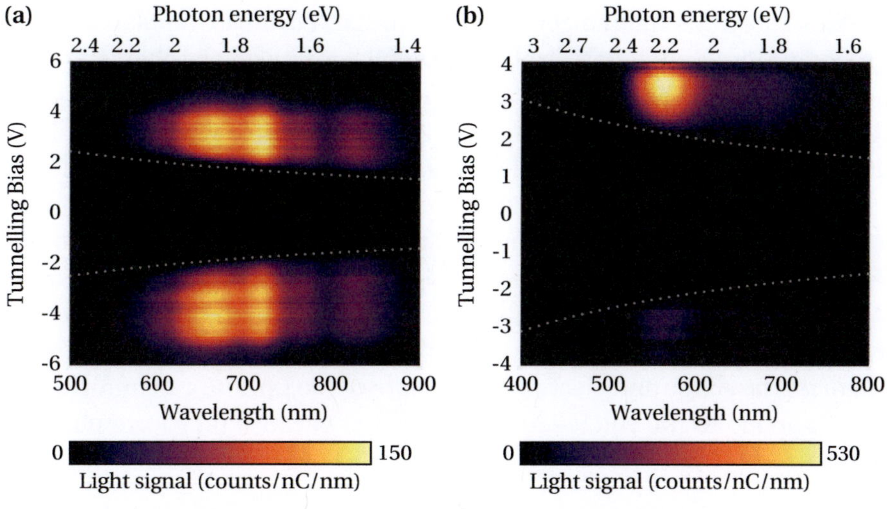

Figure 4.5: So-called *voltage maps* recorded on clean Au(111). Light intensity at different wavelengths depending on the bias voltage at a constant current of 1 nA. The dotted grey lines represent the energy conservation cut-off for one-electron processes. (a) Recorded with a blunt tip within 33 minutes, containing 93 spectra. (b) Recorded with a sharp tip exhibiting a much more well defined resonance and higher brightness. The total measurement time was 6.3 minutes for all 52 spectra.

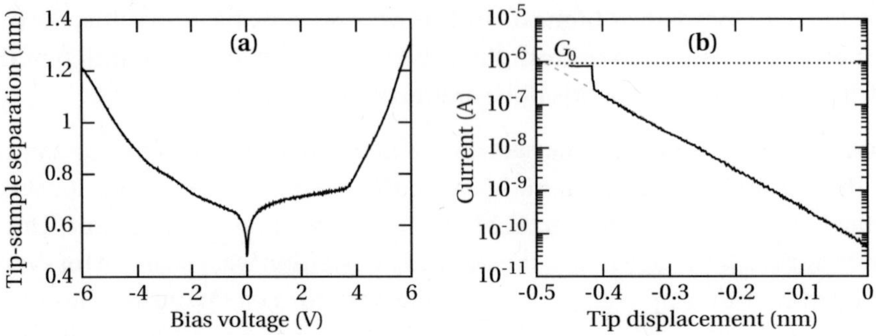

Figure 4.6: Tip-sample separation on Au(111) recorded with a tungsten tip. (a) Voltage-distance characteristic at a constant current of 50 pA. (b) Current-distance characteristics at 12 mV with exponential fit, used to determine to total offset of the curve in (a). At zero displacement the current is 50 pA like in (a). The dotted vertical line indicates the conductance quantum G_0 which corresponds to an atomic contact.

From this measurement alone, only the relative changes of the tip position can be obtained, but the absolute tip-sample separation is unknown. This can, however, be determined with a second experiment presented in Fig. 4.6(b). In this, the voltage was set to 12 mV, which corresponds to the points in (a) closest to zero bias, and the tip was approached to the surface with switched-off feedback loop. The well-known exponential increase of the tunnelling current was then measured and extrapolated to the point, where a single atomic contact is established, which is universally characterised by the conductance quantum $G_0 = 2e^2/h$ [216]. The necessary vertical tip displacement to reach this atomic contact then corresponds to the tip-sample separation at 12 mV and 50 pA and was added to the curve in Fig. 4.6(a). As a result, Fig. 4.6(a) faithfully reproduces the absolute tip-sample separation depending on the bias voltage at 50 pA and can be generalised to any current setpoint by adding the corresponding offset from Fig. 4.6(b). For example, both voltage maps in Fig. 4.5 were taken at a current setpoint of 1 nA, so that the tip was 145 pm closer to the sample than the curve in Fig. 4.6(a) indicates. However, the exact shape and offset of this curve is, among other things, dependent on the shape and electronic structure of the tip, and can unfortunately not be directly applied to the voltage maps in Fig. 4.5, because they were obtained with different tips.

It still serves to demonstrate some universal tendencies. For positive polarity, for example, the slope of the voltage-distance characteristic between 2 and 4 V is almost linear and quite flat. At this tip-sample separation, the electric field enhancement at the tunnelling junction is high, and intense light emission can be observed across a wide voltage range. At higher bias, the field emission regime is approached and the tip-sample separation then abruptly increases much more steeply, thereby reducing the enhancement of the electric field which couples the oscillation of the conduc-

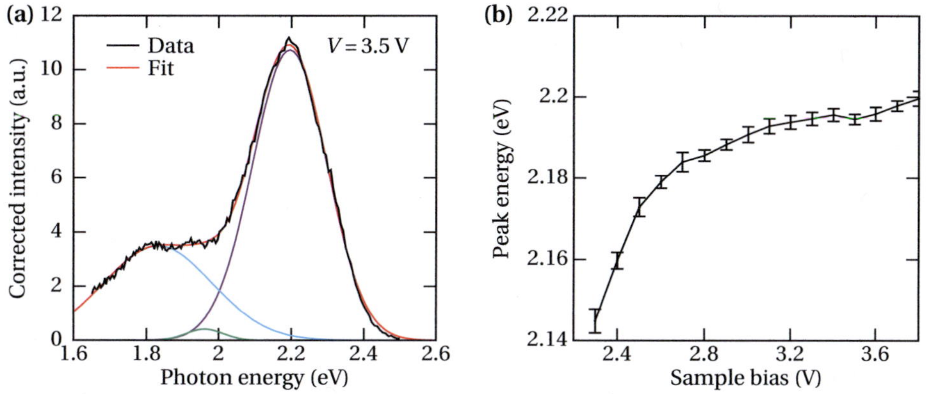

Figure 4.7: Energy shift of the main plasmonic peak in Fig. 4.5(b) depending on the tunnelling bias V. (a) Example spectrum at $V = 3.5$V. The measured light intensity was corrected for the detection response and plotted in terms of photon energy instead of wavelength. The triple Gaussian fit (red) is depicted as well as the constituent Gaussians (blue, green, purple). (b) Fitted energy of the main peak in (a) at different sample biases.

tion electrons in the tip and the sample. Such a behaviour can explain for example the sharp upper voltage boundary in the voltage map in Fig. 4.5(a) and demonstrates how sensitive the plasmonic resonance is to the width of the gap. At negative polarity, however, the tip position in Fig. 4.6 is increased much more quickly already at lower absolute values of the bias voltage. This behaviour fits to the voltage map in Fig. 4.5(b), in which the intensity at negative sample bias is very low, presumably because at voltages, that were sufficiently high for exciting the plasmons, the large tip-sample separation already caused lower field enhancement. The reason for this asymmetry lies in the tunnelling matrix element, which involves different electronic states of the tip and the sample depending on the direction of the tunnelling current, and therefore results in a different tunnelling probability depending on the bias polarity. With different tips, the voltage dependent tunnelling probability can be very different, and, as a consequence, result in a different voltage-distance characteristic at constant current. In the voltage map in Fig. 4.5(a), the sample separation at negative bias must have been much smaller than in Fig. 4.5(b), so that intense emission could also be observed at comparably high negative bias voltages.

In the voltage maps, it looks as if only the intensity of the spectra varied with the bias voltage and the shape stayed unaffected. While it is true that the number of peaks is not expected to change with increasing gap width, it was found, however, that the centre wavelength of the plasmonic resonance shifts depending on the tip-sample distance [54]. To verify this effect with the data in Fig. 4.5(b), the response corrected luminescence spectra at each bias voltage were fitted in the energy space with a triple Gaussian, which faithfully reproduces the signal, as can be seen in Fig. 4.7(a). The position of the most prominent peak can be extracted from this fit with low uncertainties and is plotted in Fig. 4.7(b). At lower voltages, the peak is visibly shifted towards lower energies, as the excitation energy is not sufficient to fully excite the resonance across its entire width. At higher voltages, the peak is fully formed, but still slowly shifted towards higher energies as the sample bias increases. This is in agreement with the prediction, that the energy of the gap plasmon is higher at larger gaps [46, 51]. The total energy shift from 2.14 to 2.2 eV in Fig. 4.7(b) corresponds to a wavelength shift of about 16 nm, which is difficult to see in Fig. 4.5(b) due to the scale.

4.2.3 Spatial resolution

The key virtue of STML is its extraordinary spatial resolution, owing to the positioning precision of the STM tip and the strong confinement of the optically active area. This is showcased in Fig. 4.8, depicting a photon map obtained close to the area shown in Fig. 4.1(a) across two step edges of a Au(111) surface. While varying the lateral position of the tip along a grid, the position of the tip was recorded as well as a complete optical spectrum for every point. This mode of operation allows to attribute variations in the light emission to features of the surface. From the recorded tip positions, a conventional STM image can be reconstructed, see Fig. 4.8(a). As the full set of data is four-dimensional, the information that can be gained from the optical channel is exemplarily demonstrated by fitting the main plasmonic resonance peak (595 – 665 nm) with a Gaussian at each point and mapping the lateral variation of the intensity and peak position in Figs. 4.8(b) and (c) respectively. In these maps, several different effects can be observed, which will be discussed in the following.

At the step edges, a decrease of intensity is observed, alongside a gradual blue-shift when approaching a descending step edge, as is clarified by the profiles in Fig. 4.8(d) that were averaged over 6 adjacent lines. Such a reduced intensity at step edges has already been observed on other closely packed noble metal surfaces like Au(111) [116], Au(110) [145], Cu(111) [29] and Ag(111) [54, 153, 184] as well as Cu_3Au in a chemically ordered phase [217], whereas on Au(110), enhanced emission was observed at step edges [55, 57]. The effect was finally attributed by Hoffmann *et al.* [55] to the LDOS at the step edge. There, a one-dimensional electronic state is formed due to electrons being trapped in their own image potential [218], similarly to the well-known formation of two-dimensional image states at metal surfaces [171]. This gives rise to a modified LDOS compared to the terrace area. In particular, this leads to fewer available final states for inelastic tunnelling on many noble metal surfaces studied in the literature, so that gap plasmons are less likely to be excited.

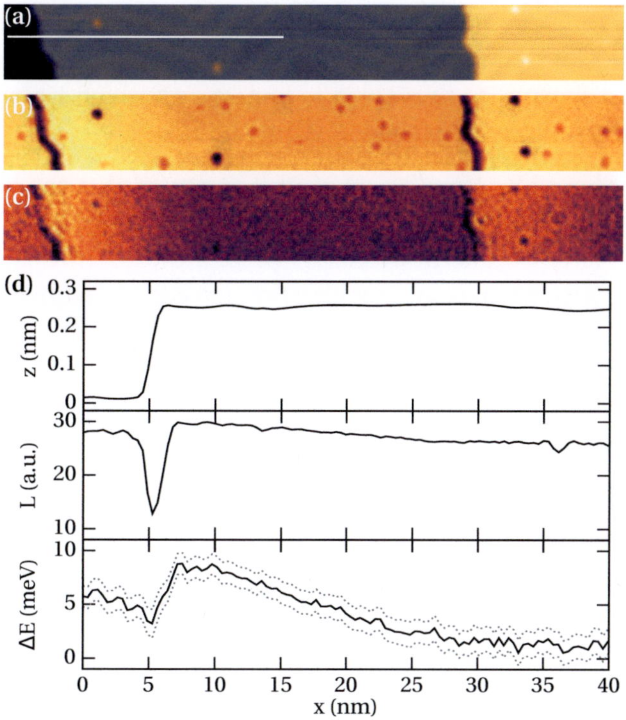

Figure 4.8: Spatially resolved photon map at a constant current of 2 nA on Au(111). A complete optical spectrum was taken at every point of a 240 × 30 grid. The applied sample bias was 2.5 V, and the exposure time was 5 s per point. (a) Topography recorded at the same time as the photon map. Subsequent lines were matched regarding their height to account for slight tip changes and vertical drift. (b) Fitted main peak intensity. (c) Shift of the fitted main peak energy relative to 1.958 eV. (d) Profiles of the topography, intensity and shift of the peak position along the white bar in (a), averaged over six adjacent rows. The grey dotted lines mark the confidence intervals found for the peak position shifts.

The blue-shift towards descending step edges apparent from Figs. 4.8(c) and (d) was also observed by Aizpurua *et al.* [54] on Ag(111) and explained by the geometric modification of the cavity caused by the step. The gap plasmon is confined to a certain area around the tunnelling junction and is, therefore, sensitive to changes of the cavity inside this confinement radius. In particular, a vertical atomic step causes a variation of the gap width, which can be treated as an increase or decrease of the average tip-sample separation experienced by the plasmon depending on whether the tip approaches an ascending or a descending step. Approaching a descending step then results in a larger effective gap width, resulting in a blue-shift of the plasmon resonance. As a consequence, the distance from the step edge, at which the blue-shift sets in, can be used to assess the plasmon confinement that was achieved in the experiment. From the bottom panel in Fig. 4.8(d), it is apparent that this confinement radius was around 20 nm for this particular mirror tip. This is comparably large, which is not surprising, considering that the mirror tip used during this experiment featured a stump instead of a sharp STM tip. It was, in fact, the tip which was already presented in Fig. 3.2(d) as an example for a failed DLW procedure.

In the model developed by Aizpurua *et al.*, it is expected that the blue-shift towards a descending step edge is complemented by a red-shift towards an ascending step-edge on the same lateral scale. In this experiment, however, this is not observed. As apparent from Fig. 4.8(c), when approaching the step edge from the left, the fitted peak energy stays constant within the accuracy of the measurement until the step edge is reached and significant vertical tip movement sets in. This might indicate a strong asymmetry of the tip with different confinement lengths to the sides. Directly at the step edge, two different effects are observed. At the spot, where the profiles for Fig. 4.8(d) were taken, the peak energy smoothly transitions to the blue-shifted value measured at the upper terrace. On most other spots in the map, however, a distinct and very narrow area of red-shifted plasmon resonances can be seen as dark contrasts that outline the step edges in Fig. 4.8(c). The observed behaviour changed throughout the measurement, so it appears that this is tip related, whose configuration has obviously changed several times during this measurement. In light of the low field confinement that was achieved with this tip, the only remaining physical quantity that influences the light emission, changes on such a short length-scale, and is very sensitive to changes of the tip is the inelastic tunnelling probability. So far, it was assumed that tunnelling into the edge state causes the excitation at all photon energies to be equally damped. This needs not be the case as the tunnelling matrix element involves the wavefunctions of the tip and the sample and is, in general, tip-dependent and energy-dependent. Therefore, different photon energies may be excited with different probability, which would shift the detected peak to another energy. Here, this conclusion cannot be regarded as definitive due to the lack of additional data, but it will be demonstrated again in Sec. 5.2.2 that variations of the energy-dependent tunnelling probability do cause such apparent peak shifts.

Apart from the dominant intensity contrast at the step edges, a gradual increase of the fitted peak height of about 10 % in total is also observed, that accompanies the blue-shift in the proximity of a descending step-edge. This increase is also apparent from isochromatic and totally integrated maps (not shown here) and is in agreement with Hoffmann and Berndt [52], albeit three times more pronounced. On the left of the profiles in Fig. 4.8(d), neither the light intensity nor the peak shift drop back to their initial values like they do on the right step edge, because there is another step edge to the left barely outside of the imaged frame.

Remarkable contrast mainly in the luminescence intensity is also caused by surface contaminants, even though some are barely visible in the STM topography in Fig. 4.8(a). Only the most prominent contaminants alter the tip-sample distance enough to affect the charge carrier coupling of the tip and the surface [134]. But the luminescence channel also provides a sensitive gauge for inelastic tunnelling processes that are obviously affected also by the much less prominent contaminants, which emphasises the need for spatial resolution in STML studies. In summary, the maps in Fig. 4.8 highlight how sensitive this mode of operation is to subtle variations of the gap cavity and the excitation probability.

4.2.4 Collection efficiency

In the black spectrum shown in Fig. 4.4, the total photon yield at the detector was 5.6×10^4 photons per nanocoulomb, and most of the mirror tips employed so far achieve a comparable brightness if worked a little bit. However, direct intensity comparison among different experiments is difficult and many publications do not state this figure at all, as the conversion efficiency of the tip is vastly unknown. Taking the response curve in Fig. 3.4 into consideration, the conversion efficiency times the collection efficiency of this particular mirror tip was 2×10^{-5}. So far, no method could be conceived which would allow to determine those two figures independently from each other. The quantum efficiency is estimated between 10^{-4} and 10^{-3}, and other groups have reported similar [47, 157] or even significantly higher [132, 152] photon yields on noble metal surfaces, despite using collection optics that cover a smaller fraction of the solid angle. So even though the detection rate that is routinely achieved with the mirror tips is higher than in many other STML setups (see Tab. 2.1) and high enough to perform experiments with high spatial, voltage or current resolution, it seems that their collection efficiency is not yet at its full potential. So optimising the DLW and metallisation procedures or exploring post-processing techniques to smoothen the mirror surface might further enhance the photon yield by as much as one or two orders of magnitude.

5 Plasmonic light emission from Co nano-islands on Cu(111)

Heteroepitaxial systems which have been investigated in STML experiments, so far, always comprised a metal in conjunction with a dielectric. Either a dielectric was used to support noble-metal grains, such that the metallic grains could be investigated individually, or, in the opposite case, noble-metal surfaces were covered with an insulating material to support molecules. Here, the first ever STML study of a heteroepitaxial system is presented in which both constituents are metallic. This is particularly interesting because the influence of the tip shape on the light emission is always dominant in STML, so that disentangling the influence of the tip and the influence of the material is very difficult, if not impossible, when two different samples are investigated consecutively and tip changes cannot be fully excluded. When the light emission is mapped on a sample that comprises two different materials, however, any relevant tip change can be excluded because it would appear as a discontinuity in the luminescence signal in between two data points. Therefore, a heteroepitaxial system allows to directly compare emission spectra from different materials which were obtained with the same tip, so that the influence of the material properties can be evaluated.

Cobalt on Cu(111) was chosen as a material system because it is well studied, and the natural occurrence of multi-layer nano-islands already at low coverage provides the opportunity to study the bare substrate and decently thick films on the same sample. That way, a higher variation of the material properties within the tunnelling junction is achieved as compared to systems with layer-by-layer growth. In the literature, Co on Cu has mainly attracted attention because of its magnetic properties. Multi-layered sandwich structures were found to exhibit alternating ferromagnetic and anti-ferromagnetic coupling between the individual Co layers depending on the thickness of the Cu spacers, and the anti-ferromagetic configurations gave rise to a giant magnetoresistance of up to 65 % at room temperature [219, 220]. The nano-islands, which form at sub-monolayer coverage, were found to be ferromagnetic with a preferred magnetisation direction perpendicular to the surface [221]. Because of its potential application, the system was then widely studied with different techniques. However, magnetic properties of the system were not studied in this work as a non-polarised Ag coated tip was used throughout the experiments and no external magnetic field can be applied to the γ-STM.

For the sake of simplicity, the measurements in this chapter were all performed at 4.4 K using neither the JT stage nor the heater. A constant and low temperature was desired for stability of the tip, low drift and to prevent surface diffusion, but the specific value of the temperature was not important. The wavelength resolution in the presented measurements was limited by the detection setup to 8 nm, which corresponds to an energy resolution of 22 meV at 650 nm. In light of that, thermal broadening of the energy of the tunnelling electrons is already negligible at 4.4 K, and the energy-resolution would not improve further by cooling to a lower temperature. Also other temperature dependent effects of plasmonic light generation, such as above-cutoff emission, can only be studied at much higher temperatures on clean metal substrates [119]. Therefore, the cryostat was operated in the simplest mode, which is at liquid helium temperature.

5.1 Growth

5.1.1 Summary of the literature

Obtaining well-defined sandwich structures requires a well controlled deposition process, and while the magnetic coupling and magnetoresistance of superlattices grown along the (100) direction were quickly well understood [222–225], contradictory results have been obtained with different deposition techniques for structures in the (111) direction [220, 226–228]. To study the growth of Co on Cu(111), different techniques have then been employed by various groups such as low-energy ion-scattering [229] and STM [230–233], in some studies also combined with low electron energy diffraction [234]. The two metals are immiscible in volume, and their lattice mismatch is 1.9 %. Therefore, this combination of materials seems suitable for heteroepitaxy and one could, at first, expect defect-free layer growth with a well-defined interface. It was, however, found that the growth of this system is quite complex because of the high difference of the surface energies of the two metals, and the morphology of grown films is very sensitive to the deposition conditions.

In the following, the growth of Co islands on Cu(111) will be explained along the lines of a recent study by Vu and Morgenstern [235] as it corroborates and summarises many earlier reports. Upon cobalt deposition onto a clean Cu(111) surface, nucleation preferentially occurs at step edges. When the temperature during the deposition is low enough or the atomic terraces of the sample are wide enough, nucleation also occurs on the terraces. At first, monolayer islands form, which are energetically unstable due to the high surface energy of Co. Bilayers are energetically more favourable because they feature a smaller Co terminated surface area. Already at 150 K, Co atoms which are adsorbed on kink sites of the island are transported to the second layer due to a low corresponding energy barrier of 0.15 eV. In islands that form at the step edges of the substrate, these kink sites are less prevalent as these islands adapt the straight shape of the Cu step edge. As a consequence, the formation of bilayer islands is more likely on islands which have formed on a terrace. For Co films deposited at room temperature, mostly triangular bilayer islands are observed on the Cu(111) terraces which reflect the anisotropy of the edge diffusion energy barriers [236]. Also a small fraction of trilayer islands forms on the terraces, whereas also a low fraction of monolayer islands can be observed at the substrate's step edges.

Most of the Co islands follow the face-centred cubic stacking of the substrate, but about 20 % grow with a stacking fault because the nucleation can initiate at two different three-fold hollow sites [230, 237]. The faulted and unfaulted stackings can be distinguished in STM topographs as they form triangles rotated by 180° with respect to each other. They also differ slightly with regard to their LDOS [237, 238].

Even though Co and Cu are immiscible in bulk, they do exhibit intermixing at surfaces. The high interface energy of these metals prevents alloying in bulk, however, the difference of the surface energies is still higher, so that it is energetically favourable to cap the cobalt islands with a layer of copper [232]. This was first discovered with low-energy ion scattering [229]. Interestingly, the copper cap is not simply added to the top of the islands, but intermixes into the top layer of the island and subsequently replaces the cobalt atoms which are then shuffled to the lower layers. This process is diffusion-limited and occurs at room temperature at a time scale of minutes to hours [229, 232]. Therefore, the deposition rate and the time to transfer the sample into the cryostat is a critical parameter.

So far, two mechanisms have been proposed to explain the dynamics of this process. One is interlayer atom exchange predicted by theoretical Monte-Carlo simulations [239, 240], the other one is trapping of diffusing copper adatoms which are sufficiently mobile to migrate onto the islands and are then incorporated into the surface [231]. Both mechanisms might occur at the same time to some extent, although only the latter can explain the observed etching of the Cu step edges [233] and the formation of vacancy islands in the substrate during and even long after the deposition of Co [230, 232]. Apparently, mass is being transported from the substrate to the islands, which explains why the volume of the islands is often larger than the deposited amount of Co if the sample was at room-temperature for a long time [239].

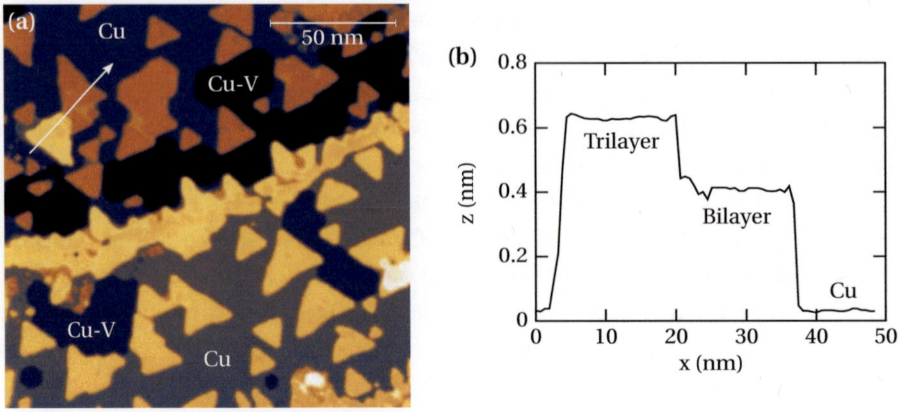

Figure 5.1: Topography of the Co/Cu(111) sample. (a) Constant current topograph taken at 1 V and 64 pA, (b) profile taken along the arrow in (a), identifying two different kinds of nano-islands.

One early STM study by Pedersen *et al.* [232] suggested that one layer of Co is buried within the Cu surface already at the beginning of the nucleation, thereby displacing the Cu atoms to the side of the newly formed island. This would mean that bilayer islands actually consist of three layers with one subsurface layer, and that all islands are decorated with Cu around their edges. Many reports followed this interpretation, however, no other study since then ever corroborated this finding. Conversely, many spatially resolved STS studies have later been published reporting on clean Co nano-islands with apparently no Cu intermixing or edge decoration [221, 238, 241, 242] which would have been easy to identify in maps of the LDOS. Also a recent, more thorough study of the nucleation process at different stages and at different temperatures did not reveal embedding of Co atoms into the Cu surface at any time [235]. Therefore, the idea of displaced Cu atoms from the topmost substrate layer generally aggregating around the edges of the islands is discarded in this thesis.

5.1.2 Surface topography

The sample used for the electroluminescence measurements was prepared by first cleaning a Cu(111) single crystal by repeated Ar^+ ion sputtering and subsequent annealing at 550 °C. The sample was then left to cool for 2 h. Then, Co was deposited from a degassed metal evaporator for 80 s at a measured ion flux current of 30 nA. The sample was quickly transferred into the STM, which was equipped with a mirror tip.

An overview of the resulting surface topography is given in Fig. 5.1 (a), showing two terraces of the Cu(111) substrate separated by a step edge running horizontally through the middle of the image. The step edge is almost fully decorated with Co agglomerates consisting of many closely spaced nucleation sites, and the terraces feature more individual triangular nano-islands, which are also often fused together. From the profiles in Fig. 5.1(b), it is apparent that most of these islands have an apparent height of 3.9 Å, which corresponds to the well-known apparent height of bilayers from the literature. Some trilayer islands are also observed at an apparent height of 6.0 Å.

On the underlying Cu substrate, extended vacancy islands, labelled Cu-V in the image, indicate that considerable copper intermixing of the cobalt islands must have occurred. This has to be attributed to the poor control of the sample temperature after the annealing. Had the sample been at room temperature during the Co deposition, the time for the preparation and the transfer into the STM should have been sufficiently short to obtain Co islands without intermixing [221, 232]. Apparently, however, letting the sample and the preparation manipulator stage cool for 2 h after the annealing was not sufficient to reach room temperature, and this intermixing process is indeed known to occur more rapidly only slightly above room temperature [229]. As a consequence, the intermixing occurred on a shorter time scale than the sample could be transferred into the cryostat. On a large overview scan covering 200 nm × 200 nm (not shown here), it was found that 48 % of the surface is covered with islands and 20 % of the surface consists of Cu vacancies.

In the following, it will be discussed where the Cu intermixing might have occurred. On the top of some nano-islands, there are depressed areas in the STM topographs, which are imaged 30 to 60 pm lower than their surroundings, as shown in Fig. 5.2(a) and the profiles in (c). Especially the irregular islands around the step edge in the top of Fig. 5.2(a) exhibit large phases with inhomogeneous contrast, as can be seen in the inset with a different false colour scale. Also, a protruding edge decoration is visible on profile 1 in Fig. 5.2(c). On a well-defined crystalline surface, such shallow height variations do probably not represent the geometric arrangement of the surface atoms but are rather a so-called *chemical contrast* due to changes of the LDOS caused by different species of atoms [243]. It is tentative to attribute these depressions to the formation of a surface alloy, which might stem from ongoing replacement of the top Co layer with Cu atoms, which froze out as soon as the sample was transferred into the cool STM. This interpretation is compatible with the topography of a second sample shown in Fig. 5.2(b), which was cooled prior to the Co deposition. This resulted in the lack of any Cu vacancies and a different growth mode, as apparent by the more irregular island shapes and the incomplete formation of bilayers. According to the literature, the islands visible in Fig. 5.2(b) likely only consist of Co and, strikingly, lack the abundant depressions visible in Fig. 5.2(a). A comparison of the apparent heights in Fig. 5.2(c) therefore confirms that said depressions in Fig. 5.2(a) are likely not Co terminated faces.

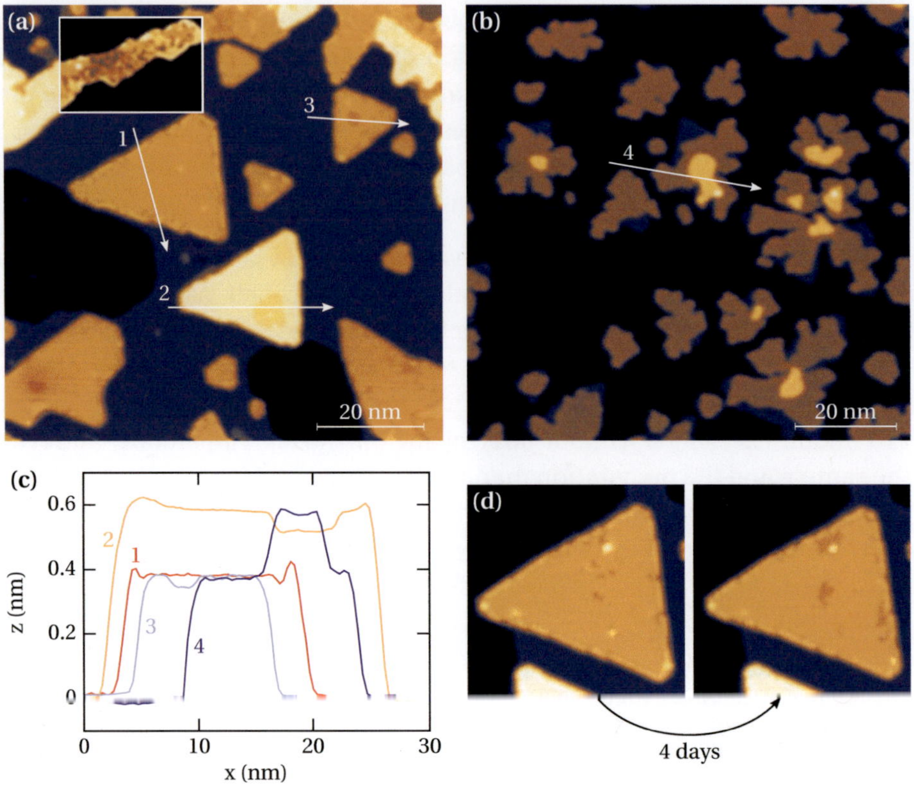

Figure 5.2: Apparent depressions on top of the nano-islands interpretated as chemical height contrast. (a) Constant current topograph taken at 1 V and 340 pA. The false colour scale in the inset was adapted to point out surface inhomogeneities. (b) Topography of a sample prepared at lower temperature, which lacks the chemical contrasts in (a). Recorded at 1 V and 100 pA. (c) profiles taken along the arrows in (a) and (b). (d) Change of apparent depressions over time as measured at similar tunnelling parameters of 100 pA on both pictures, and 1.8 V on the left and 1.5 V on the right picture.

However, in Fig. 5.2(d) it was observed that some of these dark spots rearranged over the course of several days. In between these two pictures, two photon maps have been acquired in this area, one of which will be presented in the following section. In the centre of the island, several dark spots have merged, and the overall depressed area has significantly increased. Diffusion of atoms within the surface must have been completely suppressed at 4.4 K, considering that experiments at 150 K have investigated the surface composition of this system in great detail without observing such a phase separation [232]. Local heating by the tunnelling current can also be excluded as it never exceeded 2 nA and both the thermalisation of the direct current within some 10 nm around the junction and the contribution of inelastic processes within the junction are expected to cause a temperature increase in the order of 10^{-2} K at

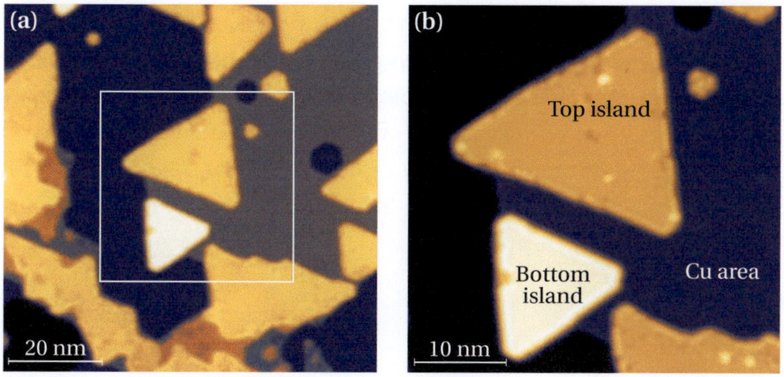

Figure 5.3: Location where the photon map was taken. (a) Overview scan including some of the surrounding area. The white square indicates the frame of the photon map and is enlarged in (b). (a) was recorded at 1 V and 64 pA, (b) at 1.5 V and 100 pA.

these tunnelling parameters [244]. Also other influences of the STM tip, such as the electric field or inelastic tunnelling processes, can be considered unlikely to induce atom-exchange hopping in a closely packed metal surface at such low temperatures. Therefore, at least some of the imaged depressions might have been caused by a layer of adsorbates, which is easily rearranged by the STM tip.

As a conclusion, it is not entirely clear where the Cu intermixing on the investigated sample occurred. The aforementioned edge decoration of the central bilayer island in Fig. 5.2(a) was attributed to Cu atoms around the edges of otherwise clean Co islands before [235], but accounts only to a small amount of Cu atoms and cannot explain the large Cu vacancies. It seems plausible that at least most of the visible depressions on the islands indicate mixed phases in the top layer, despite the finding that some of these depressions might stem from contaminants. The observation of large intermixed phases on top of the islands which have nucleated at the step edge (inset in Fig. 5.2(a)) is in agreement with a STM study in which carbon monoxide was utilised to selectively mark Co terminated faces [234]. Also, Pietzsch *et al.* [238] report intermixing primarily at the islands around the step edges. Following this interpretation, islands on the terraces without apparent depressions at the top are considered to be in a pure Co phase, and most of the intermixing indeed occurred at the irregular islands at the substrate's step edges.

However, it cannot be fully excluded that the copper intermixing might have occurred also in the seemingly clean parts of the islands without affecting the apparent height. Even though the copper vacancies on the surface amount to less than half of the area of the islands, it is still possible that a larger fraction of the islands is capped with a layer of copper.

The marked area in Fig. 5.3(a) was finally chosen for the STML measurements presented in the next section. It features two different nano-islands that showed no or as little surface inhomogeneities as possible, and can therefore be considered pure Co islands. The top island is a bilayer island with faulted stacking as determined by the direction of the triangle, and comprises a circumferential decoration, which, following the interpretation of Vu and Morgenstern [235], consists of Cu atoms. The bottom island is of trilayer height and unfaulted. Both are located at the edge of a large vacancy.

5.2 Photon map

At the scan frame depicted in Fig. 5.3, the tip position was moved along a grid of 160×160 points and an optical spectrum was recorded at each point. Muons from atmospheric cosmic showers occasionally penetrate the detector and generate large charges in the CCD, leading to erroneous peaks in single or at most a few adjacent bins. To filter these events, two spectra were recorded at each point of the grid and later automatically compared point by point to identify and discard the bins that were hit by a muon. Discarded values were then replaced with the corresponding value of the other, unaffected spectrum, as the muon events were rare enough to never affect both spectra taken at the same point in the same bin. That way, complete spectra virtually without filtering artefacts could be obtained at all points of the grid. The total exposure at each point was $2 \cdot 1.5\,\mathrm{s} = 3\,\mathrm{s}$. This resulted in a total measurement time of 29.4 h. It is noteworthy that reported long-term STML measurements so far were limited in that respect to 1 to 3 h [47, 77, 82, 86] presumably due to the limited stability of either the STM or the collection setup. Compared to other experiments, the long measurement time of the experiment presented here resulted in spectra with higher signal-to-noise ratio at high spatial resolution.

During the presented experiment, the detection setup was not yet fully optimised. The NA matching lenses were not yet installed at the time, and the spectrograph was equipped with a 150 grooves per millimetre grating, resulting in the same wavelength resolution of 8 nm discussed in Sec. 3.1.3. The achievable count rate, however, was found to be 40 % higher after the NA matcher was installed.

Fig. 5.4 shows three representative spectra obtained from the Cu substrate and both of the nano-islands at a constant current of 1.5 nA. From these, it is already apparent that the general shape of the spectrum hardly depends on the lateral tip position, but the intensity very much does. This lateral variation is demonstrated in the inset of Fig. 5.4, in which the intensity was integrated over a certain photon wavelength range for each point of the grid to generate an isochromatic map. The arrows next to the map indicate in which order the points of the grid were recorded. This is important for the drift correction discussed later, and for assessing how potential tip changes propagate through the picture. Starting in the lower left corner, all points in the respective column were subsequently recorded from bottom to top, and then the tip was reset to the bottom and the next column to the right was started.

The spectrum itself is characterised by one main resonance at 642 nm or 1.93 eV with an extended shoulder at higher wavelengths. This is the signature of a sharp tip with high plasmon confinement as will be demonstrated later by the lateral extent of the step edge effect. Significant light emission is observed between 580 and 770 nm. Similarly to the discussed spectra taken on gold samples, the dielectric function of copper restricts high emission to photon energies around 2 eV [46], which fits to the spectra in Fig. 5.4. The total detection rate was 1.88×10^5 counts per spectrum when the tip was placed above the Cu substrate. This bright signal facilitates the investigation of subtle effects as will be detailed later in this chapter. For easier comparison with other measurements, the photon counts are normalised to the tunnelling current and the exposure time, i.e., the tunnelling charge per spectrum. On the Cu substrate, the total yield was 4.17×10^4 counts per nanocoulomb. On the top and bottom islands, the yield was 1.60×10^4 and 2.26×10^4 counts per nanocoulomb, respectively.

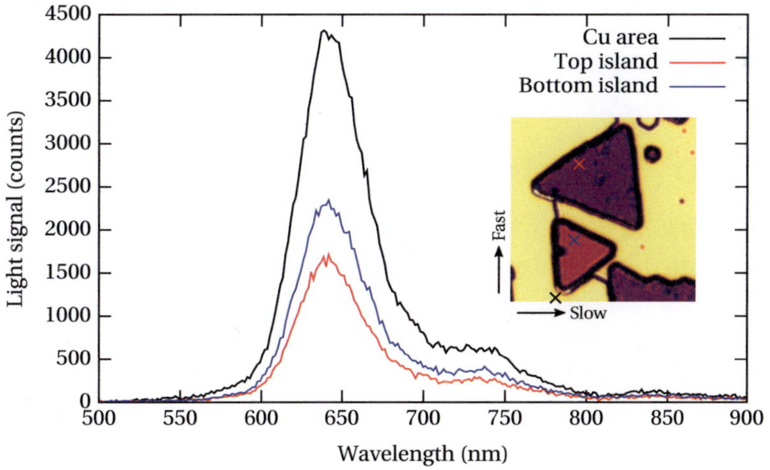

Figure 5.4: Electroluminescence spectra obtained from different locations in Fig. 5.3(b). Tunnelling parameters were 2.4 V and 1.5 nA, and the exposure was 3 s. The inset is an isochromatic map showing the lateral distribution of the integrated intensity between 635 and 645 nm. The tip positions at which the three example spectra where taken are marked with crosses of corresponding colour. The arrows next to the inset indicate the order in which the points of the grid have been recorded.

From the lateral dependence of the light emission at constant current, different effects will be demonstrated in the following. Of those, the influence of the LDOS of the final states on the inelastic tunnelling probability is by far the most dominant. It is reflected in the maps in the form of standing wave patterns, the reduced intensity directly at the step edges and is probably also responsible for the prominent intensity difference between the substrate and the islands. On the other hand, the variation of the cavity formed by the tip and the sample when moving the tip across the sample is responsible for a resonance shift at the step edges, which could be observed for different step heights for the first time. Finally, possible variations of the electromagnetic cavity caused by different dielectric functions at the surface will be discussed.

5.2.1 Friedel oscillations

Electronic states in isochromatic photon maps

In isochromatic maps of the photon intensity, standing wave patterns with a wavelength of approximately 1 nm are observed. This is exemplarily shown in Fig. 5.5, in which three areas of interest have been extracted and are displayed in the middle row. The standing waves emerge on the Cu area at step edges and surface impurities, and on the islands. This is no plasmonic feature but resembles standing waves of the electronic surface state, so-called Friedel oscillations, which form because of partial reflection of incident electron waves and interference between the incident and reflected waves. The LDOS ρ therefore exhibits oscillations close to scattering centres, which can be shown to follow the form [175]

$$\rho(E(k), x) = \rho_0\big(1 - J_0(2kx)\big) \tag{5.1}$$

for reflections at a hard wall. In this, J_0 is the zero-order spherical Bessel function and x the distance from the step edge. This is a well-known effect in STM and was observed in STS measurements of clean surfaces [173, 174] and also Co islands on Cu(111) [221, 241]. The interference pattern on the islands is additionally subject to quantum confinement due to their small size, and the wavelength can only take discrete values. For comparison, a map of differential conductance was recorded at the same location where the photon map was taken, see Fig. 5.6. The differential conductance is related to the LDOS at an energy of eV above the Fermi edge, and the standing waves imaged in this map resemble the ones seen in the isochromatic map in Fig. 5.5.

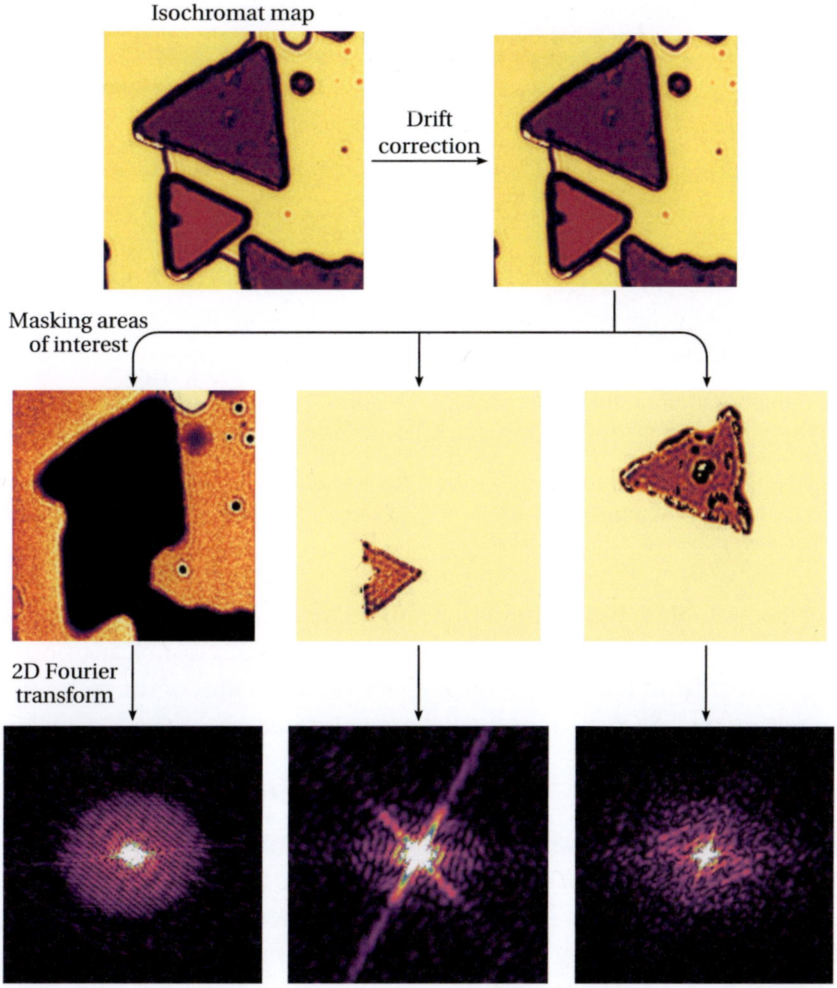

Figure 5.5: Extraction of the standing wave patterns from isochromatic maps, demonstrated for the map of the added intensity from 635 to 645 nm.

The reason why these variations of the LDOS manifest in the photon maps is that the number of available final states modifies the probability for inelastic tunnelling and, therefore, the light emission intensity [29]. Energy conservation dictates for the energies of the initial and final states of a tunnelling electron if it excites a plasmon with energy $\hbar\omega$

$$E_{\text{final}} = E_{\text{initial}} - \hbar\omega. \tag{5.2}$$

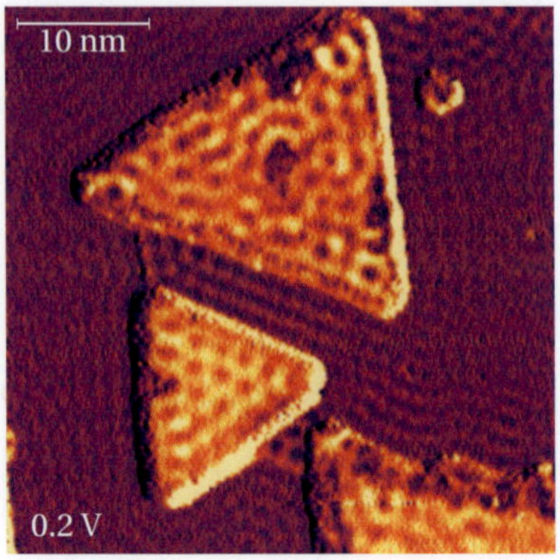

Figure 5.6: Demonstration of the Friedel oscillations using elastic STS. This map of the differential conductance was recorded with the lock-in module at a modulation voltage of 20 mV, a direct voltage bias of 0.2 V and 1 nA of tunnelling current. A Gaussian filter was applied to the image for clarity.

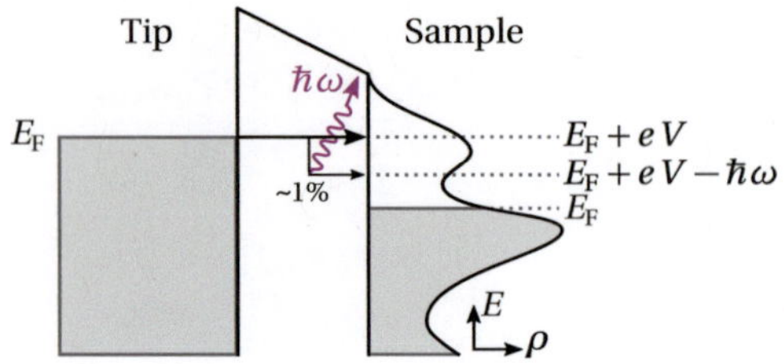

Figure 5.7: Relevant energy levels for photon emission from inelastic tunnelling at positive sample bias. The probability for inelastic tunnelling scales with the available final states $\hbar\omega$ below the elastic channel.

At positive sample bias, the emitted light intensity P at a certain photon energy — which equals the plasmon energy $\hbar\omega$ — is proportional to the inelastic portion of the tunnelling current from the tip into the respective final states of the sample, which in turn depends on the LDOS ρ of the available final states

$$P(\hbar\omega) \propto j_{\text{inel}}(E_{\text{initial}} \to E_{\text{initial}} - \hbar\omega) \propto \rho(E_{\text{initial}} - \hbar\omega), \tag{5.3}$$

whereas j_{inel} makes up only a small part of the total tunnelling current, such that variations of the branching ratio of the elastic and inelastic tunnelling channels cannot be observed in STM topographs. Assuming that only electrons at the Fermi edge of the tip contribute to the tunnelling current, ie., all tunnelling electrons are initially at

$$E_{\text{initial}} = E_{F,\text{tip}} = E_{F,\text{sample}} + eV, \tag{5.4}$$

which is the situation sketched in Fig. 5.7, then the emitted intensity at a certain photon energy directly reflects the LDOS of the available final states at a fixed energy

$$P(\hbar\omega) \propto \rho(E_{\text{final}} = E_{F,\text{sample}} + eV - \hbar\omega). \tag{5.5}$$

Therefore, the spatial distribution of the LDOS of the final states can be probed in this single measurement at a wide range of energies by rendering isochromatic maps for different photon energies. The assumption in Eq. 5.4 is unusual in STM, though, and only justified if the LDOS of the tip has a sharp maximum at the Fermi energy. Usually, it needs to be considered that also electrons below the Fermi edge contribute to the tunnelling current. Schull *et al.* [121] have already observed Friedel oscillations in STML measurements and they needed to correct the relation between the photon energy and the energy of the final state for precisely these contributions. In the following, it will be shown that this correction proves unnecessary here and the approximation in Eq. 5.5 holds well for this experiment.

Coherence length

The standing waves on the copper substrate are best observed at the upper terrace of a step edge. At the lower terrace, the first intensity maximum is obstructed by the more pronounced intensity minimum caused by the step edge effect, and the Friedel oscillations are known to be more pronounced on descending step edges [175, 245]. Therefore, a profile of the light intensity was extracted at an upper terrace from an isochromatic map shown in Fig. 5.8. Compared with the purely electronic STS measurement in Fig. 5.6, it strikes that the coherence length of the imaged standing waves is much lower in the photon maps, particularly on the Cu substrate.

For once, the Bessel function itself decays with $1/\sqrt{x}$, but the coherence length is ultimately limited by inelastic electron-electron scattering, which reduces the mean free path of the electrons in the surface state and, thus, destroys coherence. This manifests as an additional exponential damping term [245]. This effect is, however, prevalent at much higher energies. For electrons in a surface state at 1 eV above the Fermi edge, coherence lengths of 18 nm have been reported on Cu(111) [245], and the STS measurement in Fig. 5.6 and the photon maps presented here probe the surface states at much lower energy where the lifetimes are longer. Therefore, decoherence due to electron-electron scattering can be neglected in the following in light of other limiting effects.

At lower electron energies, the damping of the LDOS oscillations is usually ascribed to the limited energy resolution of the measurement, because a range of states with different wavenumbers is imaged at the same time. The coherence length scales with the wavenumber spread according to $l_c \approx \Delta k^{-1}$ [174, 175] because the different wavelength components de-phase at increasing distance from the scattering centre. This can also be modelled with an exponential damping term. The differential conductance map in Fig. 5.6 was recorded with a bias modulation amplitude of 20 mV, which determines the energy resolution of this measurement. This translates to a wavelength spread of $\Delta k = m^* \Delta E/(\hbar^2 k) \approx 0.04\,\mathrm{nm}^{-1}$ according to the dispersion relation of the copper surface state determined by Pietzsch *et al.* [221], and thus to an expected coherence length of 25 nm. A quantitive comparison will be omitted here because the probed area features too many scattering centres close by to allow the undisturbed

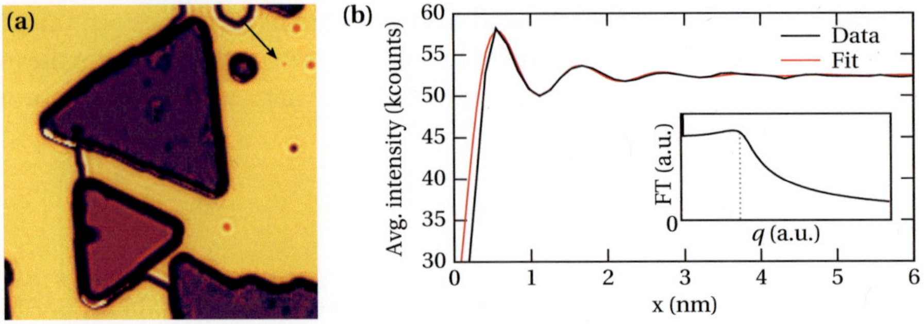

Figure 5.8: Friedel oscillations on an (a) isochromatic map integrated from 630 to 650 nm, (b) profile taken along the arrow in (a) averaged over 6 adjacent lines, fitted with Eq. 5.6. The inset illustrates the shape of the Fourier transform of these oscillations.

formation of the standing waves over such a long distance. But considering that the interference pattern forms nicely across the entire substrate area, this figure seems plausible.

In the photon map in Fig. 5.8, the energy spread is 61 meV from the photon energies over which the isochromatic map is integrated. Another 22 meV have to be added because of the detector resolution. Yet another contributing factor might be the energy distribution of the tunnelling electrons, but temperature-related smearing effects are much lower than that at 4.4 K, and the assumption in Eq. 5.4 is considered to hold precisely for now, which will be justified later. A worst-case estimate for the energy resolution therefore yields $\Delta E \approx 83\,\text{meV}$, which corresponds to a momentum spread of $\Delta k \approx 0.14\,\text{nm}^{-1}$ at the observed energy. This is similar to observations at room temperature, where the thermal energy spread is at full-width half-maximum $3.5 k_B T = 89\,\text{meV}$ [175]. This would let one expect the coherence length to be limited to $l_c \approx 7\,\text{nm}$.

This expectation is now compared with the measurement. In Fig. 5.8(b), a profile of the imaged oscillation is extracted at a spot where it is best visible and three to four wave trains can be observed. The distance dependence of the intensity is fitted according to Eq. 5.1 with an additional exponential damping term

$$I(x) = A \cdot \left(1 - J_0(2k(x - x_0)) \cdot \exp\left(-\frac{x - x_0}{l_c}\right)\right) + I_0. \tag{5.6}$$

It is noteworthy that this equation only applies to straight one-dimensional edges, and the spot chosen for the profile is not ideal due to the short length of the edge and other scatter centres close by. Still, it is the largest unobstructed descending step edge on the map and, in fact, the only spot on any map where the signal-to-noise ratio was high enough for the fitting procedure. The fit yields a damping length of merely

$l_c \approx (1.7 \pm 0.2)$nm, which is consistent with the observed standing waves around the other step edges and point scattering centres, which never exhibit more than a few wave trains. One could argue that the determined value is too small because the evaluated step edge is short and can only be approximated as a one-dimensional reflecting barrier from short distances, and looks like a point scattering source from afar. But since the obtained coherence length is similar to the step length, this correction is not substantial enough to explain the deviation from the expected 7 nm. An apparent coherence length of 1.7 nm would correspond to an energy spread of 330 meV, which is inconsistent with the previous considerations.

Apparently, either the physical coherence length of the surface states is drastically reduced due to an unaccounted for scattering process, which only affects STML experiments, or there is an additional unidentified effect which affects the apparent coherence length in the photon maps, i.e., an experimental shortcoming. A possible mechanism for a physical limitation would be inelastic electron-plasmon scattering, which is almost invisible in regular STS because the lifetime of a gap plasmon (femtosecond range [70, 71]) is much shorter than the average time difference between individual tunnelling events at currents around 1 nA (order of 100 ps). Therefore, only those electrons that excite a plasmon can also subsequently interact with it. When recording STS maps, this applies to only about 1 % of the involved electrons and would result in a minor correction to the mean relaxation time, provided the tunnelling bias is high enough to excite any plasmons, which is usually not the even case. In contrast, this inelastic fraction of the tunnelling current is the only contribution to the light signal detected in STML. Therefore, the coherence length of observed Friedel oscillations might be a probe for electron-plasmon scattering in an STML measurement, but not in STS.

This hypothesis appears unlikely, though. First, the standing wave patterns of the surface state have only been reported in one other STML study, so far [121]. In this, the authors investigated a Au(111) surface and observed long-range oscillations of the LDOS with low damping. Considering their similar electronic structure, it seems unreasonable to assume that a scattering process, which is dominant on Cu(111), cannot be observed on Au(111). Second, the interference pattern exhibits confinement behaviour on the Co nano-islands leading to a discrete wavenumber spectrum, which is shown later. This indicates that the physical coherence length of the involved electronic states is actually at least in the order of the island's size, so a dominant electron-plasmon scattering process can also not be confirmed for the Co islands. Thus, it seems more likely that there is a yet unknown factor which affects the apparent coherence length on the Cu(111) surface. Two major differences between this measurement and the one by Schull *et al.* are a higher bias voltage (2.4 V vs. 1.7 V) and a significantly lower current (1.5 nA vs. 300 nA), whereas the current should be irrelevant in this regard. The lifetime of the final state should also not depend on the energy of the initial state, thus, the different bias voltages can also not explain this discrepancy.

Dispersion relations

On a single isochromatic map, the energy of the imaged states is known from Eq. 5.5, and the wavenumber can be determined from the periodicity of the imaged interference patterns. This yields one point of the dispersion relation $E(|\vec{k}|)$ of the involved surface states. By applying this procedure to different maps which integrate over different sections of the optical spectrum, the dispersion relation can be extracted within a certain energy range. In the following, it will be explained how the wavenumber is extracted from the isochromatic maps. Then, the obtained dispersion will be compared to STS measurements from the literature.

As already indicated in the bottom row of Fig. 5.5, the 2D Fourier transform (FT) of the extracted regions of interest is used to determine the periodicity of the Friedel oscillations, which is a common technique [246]. The advantage of using the FT over extracting profiles such as in Fig. 5.8 and fitting them with an appropriate function, is that the FT evaluates the entire picture and, as a consequence, provides a much higher signal-to-noise ratio. This is an important aspect because the fitting procedure requires a good signal-to-noise ratio and was only successful in the area extracted in Fig. 5.8 on a map of the brightest part of the spectrum. The FT, in contrast, also yields plausible values for the wavenumbers on the maps featuring low intensity, as will be shown later. Note that the term *wavenumber* was not used in a precisely defined fashion so far. In the following, wavenumbers q will refer to the inverse of the wavelength $q = \lambda^{-1}$ and angular wave numbers will be denoted as $k = 2\pi q$.

Before extracting any wavenumbers, the recorded maps first need to be corrected for the small, but noticeable drift which occurred during the measurement. As the actual drift trajectory cannot be recorded during such an experiment, the next best approximation is to assume a constant drift velocity $\vec{v}_{\text{drift}}$ such that drift corrected image resembles a regular STM image of this area. The best match was achieved by assuming a drift velocity of

$$\vec{v}_{\text{drift}} = \begin{pmatrix} -82.8 \\ -28.8 \end{pmatrix} \text{pm}\,\text{h}^{-1}. \tag{5.7}$$

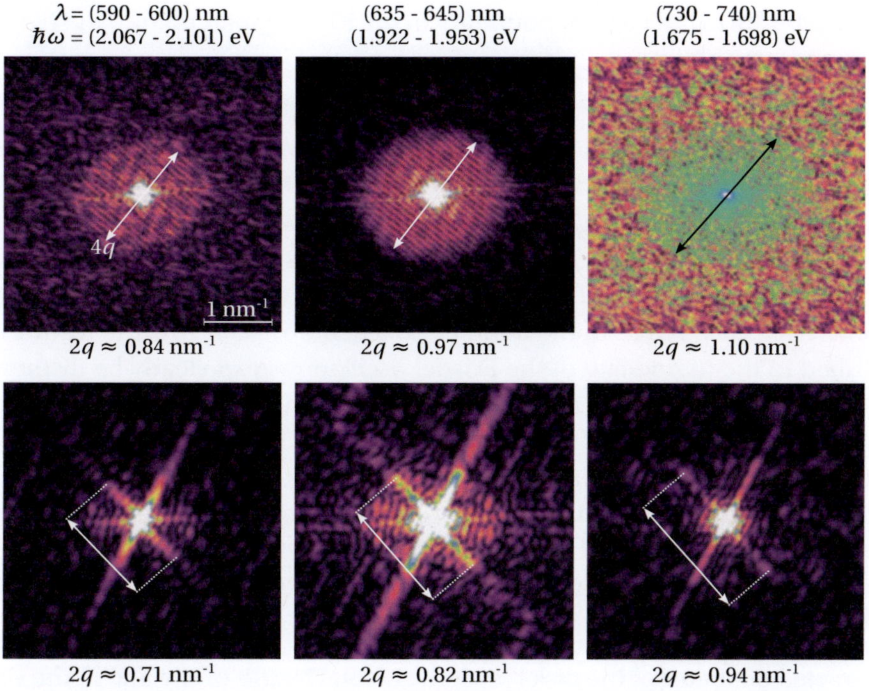

Figure 5.9: Extracted wavenumbers from Fourier transformed isochromatic maps integrating over different photon wavelength ranges. The top row represents the transform of the masked Cu area, the bottom row contains the FT maps of the lower Co island, see Fig. 5.5.

This is a little higher than the determined long-term drift in Sec. 4.1.3, but still only amounts to a total position shift of 2.6 nm within the entire measurement time. The original and corrected images are put side by side in the top row of Fig. 5.5, and after cropping the corrected image to a rectangular shape, the change is barely noticeable.

It is necessary to identify the expected feature in the FT which corresponds to the periodicity of the observed interference patterns. For this purpose, the FT of Eq. 5.6 is included in the inset of Fig. 5.8(b), which exhibits a counter-intuitive behaviour. Instead of a pronounced peak at the wavenumber of interest, this FT is almost flat at all wavenumbers below it and then drops. Generalised to two dimensions with waves along all directions — like point scattering or a mix of edge scattering in different directions across the image — the FT is expected to show a solid disk. This matches the observed pattern in the FT of the light intensity on the copper area, see the top row of Fig. 5.9. The lower island comprises a resonator in which the interference pattern exhibits wave trains parallel to the island edges. Therefore, only lines in three distinct directions are observed in the FT, see the bottom row of Fig. 5.9. The inhomogeneities on the top island, however, provide so many scattering centres that the wave pattern is too disturbed to extract a clear boundary in the FT, as apparent from the right image in the bottom row of Fig. 5.5.

For extracting the dispersion relation, 26 isochromatic maps were rendered which cover the wavelength range from 590 to 740 nm. The interval between each map is 5 nm up to 700 nm and 10 nm from then on, and each map integrates over a wavelength range of 10 nm. For each map, the same mask was used to cut out the Cu area and lower Co island, and their 2D FT was calculated using the software *Gwyddion.* In Fig. 5.9, examples for these FT maps are presented at the very beginning and end of the probed range as well as in the middle. This demonstrates that the features in the FT ascribed to the periodicity of the Friedel oscillations can clearly be distinguished throughout the probed range, and it is apparent how the disk respectively the lines on the FT maps expand with decreasing photon energy. Of the three lines visible on the FT of the Co island, only the line starting at lower right going to upper left was evaluated, see the inserted arrows. This is because the horizontal line is poorly pronounced due to a disturbance of the left edge of the island, and the line going from lower left to upper right overlaps with an artefact caused by the shape of the cut-out, and is therefore difficult to gauge.

The extracted diameter of the disk respectively total length of the line in the FT maps correspond to $4q$. As the FT is symmetric with respect to the origin in the centre of the FT map, the periodicity of the wave patterns is reflected by the radius of the disk and half of the length of the line, so $2q$. The reason for the remaining factor two is that the LDOS is proportional to the square of the wavefunction $|\Psi|^2$. Ψ oscillates with q and, therefore, the LDOS oscillates with $2q$ and the values given in Fig. 5.9 are given with respect to that. It was verified that this technique is consistent with the determined wave number from the fit in Fig. 5.8(b).

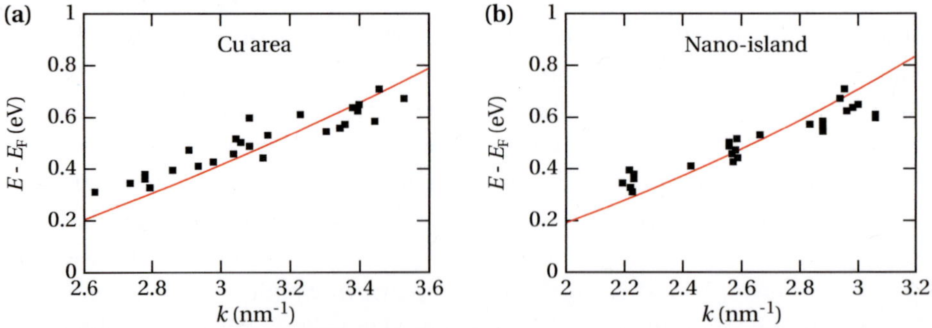

Figure 5.10: Extracted dispersion relations of the surface state on the Cu substrate and the bottom trilayer Co island (black squares) compared to the published dispersions by Pietzsch *et al.* [221] (red lines).

In the literature, it is common to specify the dispersion in terms of angular wavenumbers k. So in Fig. 5.10, the extracted dispersion is converted to those and compared with the dispersion relations reported by Pietzsch *et al.* [221] as measured by STS. Note that the reference curve for the Co island was measured on a bilayer island, whereas the dispersion presented here was obtained from a trilayer island. The only available reference curve of trilayer islands from Diekhöner *et al.* [241] indicates a small shift of the dispersion to higher energies compared to bilayer islands, but the data presented in this publication exhibit high spread and poor statistics, and are, therefore, not considered.

The surface state dispersion on the island shows a discrete spectrum with essentially only three allowed wavenumbers within the probed energy range due to the small size of the island. Pietzsch *et al.* investigated a considerably larger island and, thus, obtained smaller steps in their dispersion. Apart from this obvious detail, the values from the luminescence measurement are in remarkable agreement with the reference curves for both the Cu area and the Co island. This finally serves to justify the earlier approximation in Eq. 5.4, namely that the initial state of the tunnelling electrons is narrowly distributed around the Fermi edge of the tip due to a lucky circumstance. Schull *et al.* [121] needed to account for significant contributions of electrons below the Fermi edge to understand their extracted dispersion. This correction is substantial and seems to be inapplicable to the data presented here.

5.2.2 Step edge effects

In the following, a different approach will be followed to evaluate the lateral variation of the luminescence signal. The light signals obtained from each point of the grid were corrected for the detector response and fitted with a triple-Gaussian in order to obtain maps of fitted parameters, in analogy to the procedure in Sec. 4.2.3. As demonstrated in Fig. 5.11, this function faithfully reproduces the curve. However, a fully parametrised triple-Gaussian fit comprises many, partially correlated degrees of freedom. This applies particularly to the small middle peak in Fig. 5.11, which is not well separated from the main peak. In the fit, the position and width of the middle peak are therefore correlated with the position of the main peak, and when fitting a noisy curve, nearly degenerate solutions to the minimisation problem will emerge with different values of the correlated parameters. Therefore, meaningful values cannot be determined for all of those parameters from noisy curves at the same time. A simple workaround to this problem is to select the parameter with the most prominent influence on the curve as a figure of interest, and fix other correlated parameters with minor influence on the curve during the fit. Accordingly, the position and width of the small

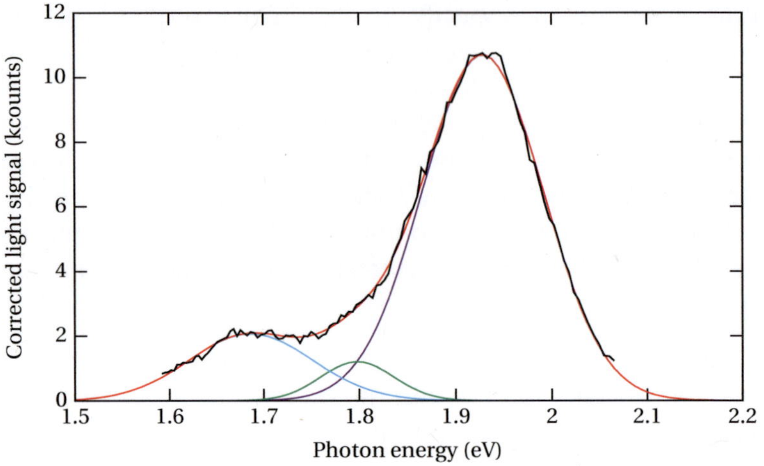

Figure 5.11: Corrected light emission spectrum recorded on the uncovered copper area (black line) with triple-Gaussian fit (red line) and the constituent single Gaussians (blue, green, and purple lines).

middle peak in Fig. 5.11 were fixed at the same, reasonable values during the fit procedures of all obtained spectra, in an attempt to extract the shift of the main peak as precisely as possible. As a result, the uncertainties for the extracted main peak positions — as calculated by the used *SciPy* routine — were between 0.3 and 0.4 meV on the areas of the substrate and the nano-islands which were sufficiently far away from the step edges, and around 1 meV at the step edges. It shall be noted that the resolution of the detection setup is far worse than that, in the sense how close two individual peaks can be and still be resolved separately. But the position of a single peak can still be determined much more precisely if it can be fitted with an appropriate function and the signal-to-noise ratio is sufficiently high.

The result of the fitting procedure is presented in Fig. 5.12 in the form of maps for the main resonance's centre energy and intensity. Two artefacts are apparent in those maps and have to be addressed first. At the spots marked by the red circles in Fig. 5.12(a), the intensity directly at the step edge is up to four times higher than on the Cu area. This was caused by an instability of the feedback loop during the measurement which resulted in a very high average tunnelling current when passing over step edges of four atomic layers in a certain direction. And the white arrow in Fig. 5.12(b) indicates a minute tip change which is characterised by the fact that all data points recorded afterwards exhibit a slightly lower peak energy, resulting in a slim stripe on the left of the maps which was omitted for all following analyses. This demonstrates how easily tip changes are identified in such maps, and no other tip change could be found in them.

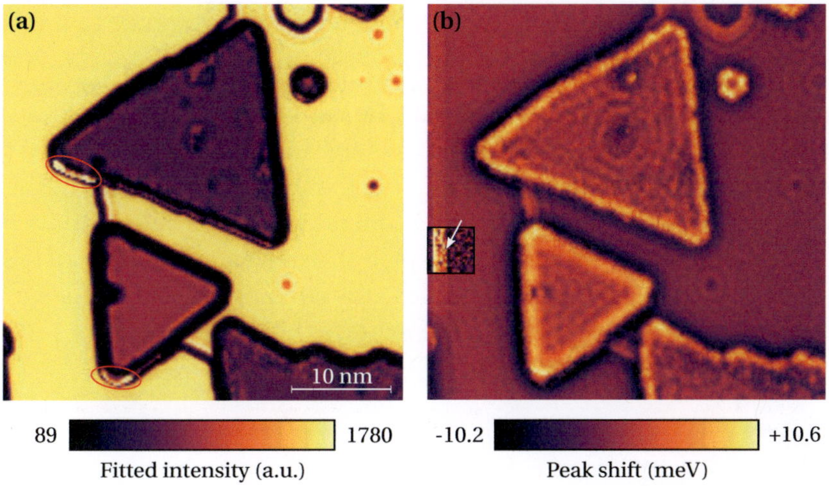

Figure 5.12: Maps of the fitted parameters for the main peak's (a) intensity and (b) energy shift relative to 1.93 eV. The red circles in (a) mark spots of irregularly high intensity caused by a instability of the feedback loop. The white arrow in (b) indicates a minute tip change, which is clarified by applying a different contrast to the inset area.

Interestingly, the Friedel oscillations are well visible in the map of the peak shift, which also demonstrates the sensitivity of this approach. This is not to be understood as a shift of the electromagnetic resonance of the gap cavity because that is only affected by the average charge carrier density within the plasmonic confinement range, which is larger than the variations of the LDOS. Also, the imaged surface state is *unoccupied* and does not even affect the charge carrier density. The origin of this contrast is rather that different sections of the optical spectrum are excited with different probability depending on the LDOS directly at the apex of the tip which causes an apparent peak shift as well.

The location where the photon map was taken provides a unique opportunity to test different ideas from the literature regarding the influence of step edges on the light emission. Within one map, there are step edges with one, two, three and four atomic layer heights as indicated by the arrows in Fig. 5.13(a). As already mentioned in Sec. 4.2.3, Aizpurua *et al.* [54] have observed a blue-shift when approaching a descending step edge, and a red-shift once the tip has passed the edge and is located on the lower terrace. Their interpretation is sketched in Fig. 5.13(c). The highly confined tunnelling current $\vec{j}$ determines the vertical tip position but the electric field $\vec{E}(t)$ of the plasmon is less confined and, as soon as the tip is placed close to a step edge, stretches over the edge and experiences an average increase of the gap width. Analogously, when the tip is placed at the lower terrace close to the edge, the averaged gap as experienced by the plasmon is decreased. It can be demonstrated here for the first time that this effect is amplified for multi-atomic step edges.

The profiles in Fig. 5.13(b) where taken along the same lines in the maps of the topography, peak position and fitted intensity, and aligned such that $x = 0$ indicates the onset of significant vertical downwards movement of the tip. The vertical tip position in the upper panel was offset for clarity, such that the middle of the step is at $\Delta z = 0$ for all profiles. Directly in front of the edge, a small protrusion can be seen in all topography profiles, indicating the edge states [55]. Due to the curvature of the tip, higher steps are imaged wider so that the lower terrace is reached at higher distances x.

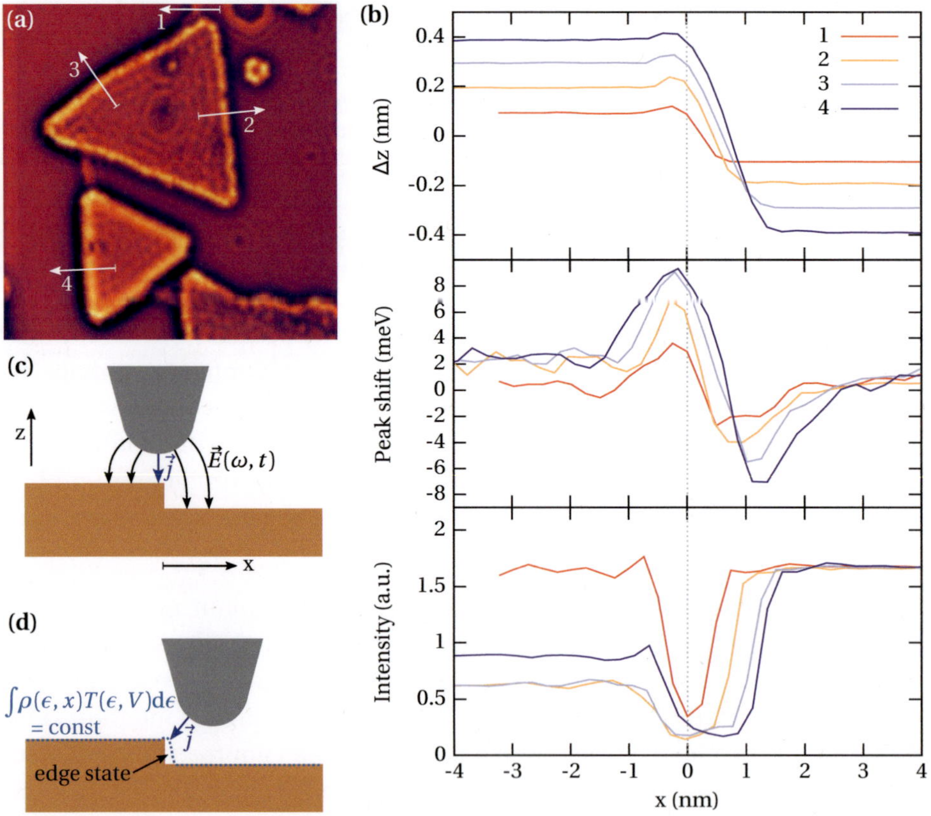

Figure 5.13: Progression of the main resonance shift and emission intensity on step edges of different heights. (a) Peak shift map from Fig. 5.12(b) and arrows indicating where the profiles for topography, peak shifts and emission intensity in (b) were taken. (c) Illustration of the cavity geometry when the tip is placed at the top of a descending step edge. (d) Spatial distribution of the integrated LDOS at a step edge including the edge image state causing the decrease of light emission.

The progression from blue-shift to red-shift upon passing the step edge is reproduced in the middle panel. It is evident how these shifts are amplified when probing higher steps, corroborating the interpretation of geometrically altering the cavity. Also, the maxima of the blue-shift precisely coincide with the maximum of vertical tip displacement due to the edge state for all profiles because the overall tip-sample separation is highest at these points. The maxima of the red-shift, in turn, are laterally shifted for the different profiles according to the delayed arrival of the tip at the lower terrace. In between, highly localised dips were visible in the map taken on Au(111) discussed in Sec. 4.2.3, which are not present in Fig. 5.13(b), so that the profiles presented here reproduce the expected behaviour of the cavity with no additional influence caused by the tunnelling probability.

From the distance from the step edge at which the blue-shift sets in, it can be estimated that the gap plasmon was confined to a radius of approximately 1.5 nm in this experiment. This is an exceptionally strong confinement and comparable to the one reported in [54], which finally demonstrates the usability of the DLW mirror tips also in this aspect.

Another well-known effect related to step edges is the drastically reduced emission. Two explanations for that have been discussed in the literature, so far. The tunnelling current always flows where the tip and the sample are closest to each other, and on a step edge, this is between the side of the tip apex and the side of the step edge. Therefore, Berndt and Gimzewski [29] concluded that the direction of the tunnelling current is not aligned to the oscillating dipole mode of the plasmon if the tip is placed above a step edge which reduces the excitation probability. As already mentioned in Sec. 4.2.3, this interpretation was later discarded and the reduced intensity was finally attributed to the LDOS of the available final states in the step edge image state [55]. The edge state often provides a low LDOS at low energies — which serve as final states for the inelastic tunnelling — and a higher LDOS at higher energies, so that the edge appears higher in STM topographs when scanning with high bias voltages. This explanation also allowed to reproduce the asymmetry of the observed intensity trough in line scans with high spatial resolution. This asymmetry is also visible in the bottom panel of Fig. 5.13(b), where the slope on the left side is smaller than on the right side of the dips. It is also apparent that the intensity minimum does not scale with the step height, which rules out any geometrical interpretation. Also, the intensity already decreases when the tip is placed directly above the edge state as indicated by the bulge in the Δz profile in front of the descent, in agreement with earlier findings [55]. Here, the direction of the tunnelling current is still normal to the surface which permits to discard the idea that the alignment of the tunnelling current and the plasmon is the dominant factor for the reduced emission intensity.

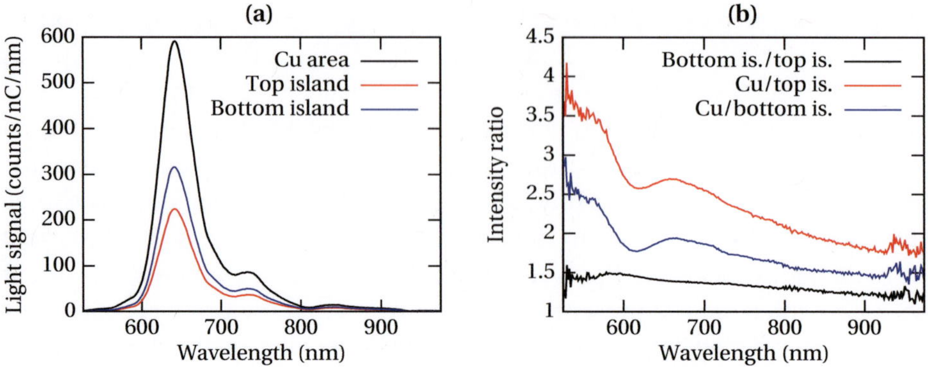

Figure 5.14: Comparison of averaged electroluminescence spectra obtained from the Cu area and both islands. (a) Averaged spectra and (b) wavelength-dependent ratios of the spectra in (a).

Another observation in the lower panel in Fig. 5.13(b) is that the trough in the intensity profile is wider for higher step edges. This does not mean that the edge state protrudes further from the step for higher steps but is again explained by the curvature of the tip. Even though this is not as important for the plasmon excitation, the tunnelling current *does* in fact flow between the side of the tip and the side of the step until the tip reaches the lower terrace. This situation is sketched in Fig. 5.13(d), and straighforwardly explains how the curvature of the tip and the height of the step determine the distance from the step along which the tunnelling current flows into the edge state.

In conclusion, the interpretations of Hoffmann *et al.* [47, 55] regarding the influence of step edges on both, the resonance energy and emission intensity, also hold well when applied to step edges of different heights.

5.2.3 Material dependent intensity

The main purpose of choosing a heteroepitaxial system was to investigate the material influence on the light emission, and, indeed, the most obvious effect apparent from the intensity maps is the intensity contrast between the Cu area and the Co nano-islands. To quantify this contrast, all spectra recorded on the Cu area and the islands were separately added, not including any spectra taken close to step edges or the contaminations on the top island. Both resulting spectra which were integrated over the area of the islands contain several million photon counts per bin around the intensity maximum, and the integrated spectrum of the Cu area features at the maximum 4.5×10^7 photon counts *per bin*. Dividing the added spectra by the number of respectively included data points yields the averaged spectra shown in Fig. 5.14(a), which exhibit virtually

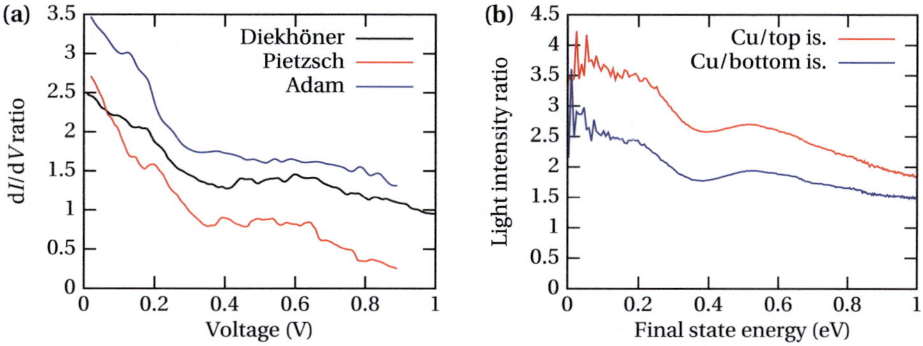

Figure 5.15: Comparison of reported $\mathrm{d}I/\mathrm{d}V$ ratios and the observed intensity ratios. (a) Ratio of the differential conductances on the copper substrates and bilayer cobalt islands taken from the works of Diekhöner *et al.* [241], Pietzsch *et al.* [238], and Adam [247], (b) extracted light intensity ratios from Fig. 5.14(b) assigned to the energies of the final states of the inelastically tunnelling electrons according to Eq. 5.5. The energy is given with respect to the Fermi edge.

no noise due to the high number of detections taken into account, and allow precise quantitative comparison. In Fig. 5.14(b), the averaged intensities from the different areas are divided by each other for each photon wavelength. The intensity ratio of the spectra from the two Co islands (black curve) is fairly independent of the wavelength, starting around 1.5 at low wavelengths and almost linearly sloping to 1.2 at high wavelengths. In contrast, the ratios between the Cu spectra and the spectra taken on the islands feature a pronounced wavelength dependence (red and blue curves). At low wavelengths, the emission from the Cu surface is ~2.5 and ~3.5 times higher compared to the respective islands. The ratio then features a local minimum at 616 nm followed by a broad local maximum at 660 nm. At higher emission wavelengths, the contrast between the various areas is lowest.

Considering the already mentioned influence of the LDOS of the final states on the light emission, it is natural to also suspect it to affect the wavelength dependence of the intensity contrast. To check this idea, the light intensity ratios at different wavelengths can be compared to energy-dependent tunnelling probabilities as determined by STS. If the tunnelling probability into a certain end state is, for example, twice as high on the Cu surface compared to a Co island, then the excitation probability for the corresponding photon energy would follow the same ratio. Unfortunately, no spectra of the differential conductance could be recorded on this sample before the tip configuration changed and the tip later ultimately failed.

However, owing to the popularity of the system, there are published STS spectra which can be evaluated for comparison. The differential conductances obtained from the Cu substrate and Co islands were extracted from the works of Diekhöner *et al.* [241],

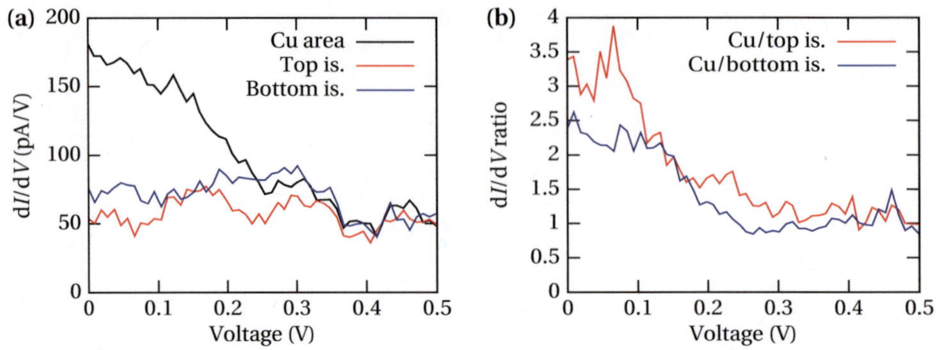

Figure 5.16: Differential conductances with the STM tip placed on top of the different areas, obtained with the lock-in at a modulation amplitude of 40 mV. (a) STS spectra, (b) energy-dependent ratio of the curves in (a) for comparison with Fig. 5.15(b).

Pietzsch *et al.* [238] and Adam [247], divided by each other and reproduced in Fig. 5.15(a). All curves differ regarding absolute numbers, which indicates that the influence of the tip's electronic structure on the tunnelling matrix element is not negligible even after division. This makes a quantitative comparison between different experiments difficult, even more so considering that STS spectra reproduce the *elastic* tunnelling matrix element, whereas the light emission depends on inelastic transitions, and the matrix elements for those are not necessarily the same. And so far, it has never been tried to quantitatively ascribe the STML contrast of any heterogeneous system by comparison to STS spectra. Nevertheless, one can recognise a similar overall shape on all curves in Fig. 5.15(a) like in the observed intensity ratios in Fig. 5.14(b). To facilitate direct comparison, the photon wavelengths in Fig. 5.14(b) were translated to the energy of the final states according to Eq. 5.5 and plotted again in Fig. 5.15(b). Strikingly, the energies of the local minima and maxima of all presented curves are similar, even though the local minima in the dI/dV ratios are less pronounced. Therefore, it is tentative to ascribe at least the overall shape of the emission intensity ratios to the differing tunnelling probabilities. The absolute values of the dI/dV ratios are, however, lower than the emission intensity ratios involving the bilayer island (red curve in Fig. 5.15(b)).

In the literature, only STS spectra taken on bilayer islands have been published, so it is unclear for now if the intensity difference between the bi- and the trilayer island is also caused by the tunnelling probability. Therefore, differential conductance spectra will be consulted in the following which were acquired from the spot where the photon map was taken — but after the tip has changed. Three spectra were recorded with the tip placed above the Cu substrate and both islands. At each of the three tip locations, the tip was stabilised at a bias voltage of 2.5 V and a tunnelling current of 2 nA, which is close to the tunnelling parameters used for the photon map. Then, the feedback loop was switched off and the bias was varied while the tunnelling current and output of the lock-in module were recorded. The latter is plotted in Fig. 5.16(a). As the electronic structure of the tip plays an essential role in the tunnelling matrix element, differential conductance spectra taken before and after a random tip change may look entirely different. Therefore, it will be impossible to draw definitive conclusions in the following. However, these STS spectra demonstrate the *possibility* that the energy dependence of the tunnelling probability might have actually resembled the curve in Fig. 5.15(b). At voltages above 350 mV, the measured differential conductance in Fig. 5.16(a) is the same for all probed areas, in contrast to the published spectra. This indicates that these spectra are dominated by the electronic structure of the tip at higher energies. At lower bias voltages, however, the differential conductances taken on the Cu area and the top bilayer island (black and red curves in Fig. 5.16) *do* roughly resemble the spectra from the literature within the achieved signal-to-noise ratio. Thus, only the low-bias section of Fig. 5.16(a) will be considered for further conclusions. There, the differential conductance above the trilayer island is about 30 % higher compared to the bilayer island's spectrum, which is actually in rough agreement with the observed light emission ratio in Fig. 5.14(b). Another striking observation is that the $\mathrm{d}I/\mathrm{d}V$ ratios obtained with the mirror tip in Fig. 5.16(b) quantitatively agree with the observed intensity ratio in Fig. 5.15(b) at low energies. This demonstrates that the ratios observed in the intensity contrast may directly reproduce the ratios of tunnelling probabilities. Therefore, it is hereby proposed that the observed light intensity ratio in Fig. 5.15(a) is for the most part caused by variations of the inelastic tunnelling probability. Accordingly, it is to be expected that the light intensity contrast between the Cu and Co areas is tip dependent, which might be confirmed in future experiments.

Still, other effects might contribute to the intensity contrast. For example, it also needs to be considered that the tip-sample separation in the constant current mode is generally not the same on different surfaces. An experiment to determine the tip distance like in Sec. 4.2.2 could not be performed with this tip, but the inter-atomic distances of the Co layers within the nano-islands as measured by low electron energy diffraction can be compared with the apparent height in STM topographs. The results of Figuera *et al.* [234] indicated a layer distance of 2.07 Å, such that the geometric height of the two islands amounts to 4.14 Å and 6.21 Å. The apparent height of the islands in STM is 3.9 Å and 6.0 Å. Therefore, it can be concluded that the tip is approximately 0.2 to 0.25 Å closer to the surface when it is placed above the islands. This increases the electric field enhancement and, therefore, the charge carrier coupling, leading to an increase of the emission and a red-shift. Berndt *et el.* [30] and Aizpurua *at al.* [54] have investigated the distance dependence of the photon yield on Cu(111) and the resonance shift on Au(111), respectively. Applying their results to a decrease of the tip-distance of 0.25 Å yields an expected intensity increase of 1.6 %, which can be considered negligible, and a red-shift of 1.5 meV, which will be discussed later.

So far, the dielectric function of the involved materials has not been addressed. It cannot be expected that the dielectric function of a few atomic Co layers matches the bulk behaviour, so first one would need to know the dielectric function of the thin film, which is not accessible in STM. However, one figure closely related to the dielectric function is the charge carrier density, and this is accessible at the surface by STS, at least in a comparative fashion. Only electrons within an interval of $\pm k_B T$ around the Fermi edge can participate in collective oscillations. Therefore, the relevant charge carrier density is the integrated LDOS around the Fermi edge, which can be qualitatively extracted from Fig. 5.16(a) at zero bias. It is apparent that this is significantly higher on the Cu area compared to the islands, and also a little higher on the trilayer island as compared to the bilayer island. So it can be concluded that there is a material related variation of the dielectric function, and the cavity formed by the tip and the Cu substrate is not electromagnetically equivalent to the cavity formed by the tip and the nano-islands. However, it is unknown how much the change of charge carrier density directly at the surface affects the cavity properties. The essential question is how deep the electric field of the gap plasmon penetrates into the sample and, thus, how much of the underlying Cu substrate partakes in the collective electron oscillation if the tip is placed above a nano-island. The skin depth of regular SPPs of a planar surface is determined by the electrons' capability to shield the electric field, and is in the order of several 10 nm [248]. Theoretical predictions for the vertical confinement in STML caused by the tip have only be published for rather blunt tips, indicating a vertical extent of around 10 nm [249]. It is unclear for now if this is applicable to a sharp tip like it was used for the photon map.

If the nano-islands make up a significant fraction of the total material that partakes in the charge carrier oscillation, one would expect the resonance energy to shift. In particular, a lower charge carrier concentration n leads to a decrease of the bulk plasma frequency $\omega_P = \sqrt{ne^2/(m\epsilon_0)}$, which would also decrease the gap plasmon frequency according to Eq. 2.14. On the Co islands, therefore, a red-shift would be expected adding to the small expected red-shift from the presumably reduced tip-sample separation.

Evaluating the fitted peak positions in Fig. 5.12(b) across the three different areas leads to the distribution histogram in Fig. 5.17, which reveals that no such red-shift is observed. First of all, the observed shifts are so small that they are at the verge of significance. The width of the individual distributions is largely due to the contrast from the Friedel oscillations, and all distributions overlap. The distribution centres are less than 1 meV apart, which is close to the determined fit uncertainty of 0.3 to 0.4 meV.

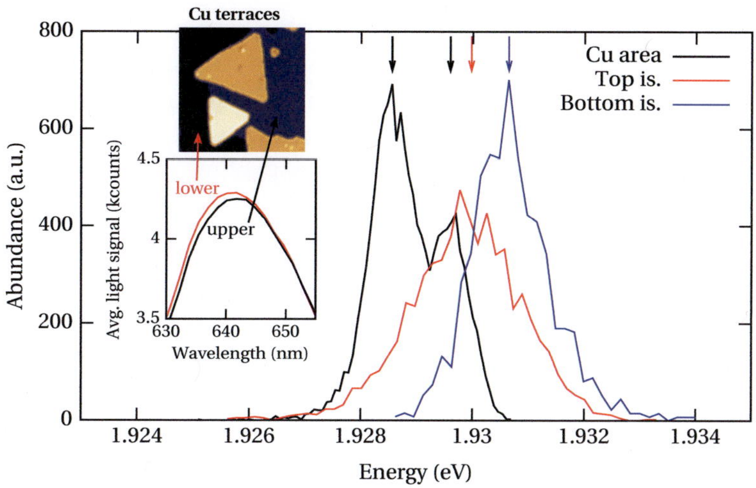

Figure 5.17: Abundance distribution of fitted peak positions in Fig. 5.12(b) on the different areas. The inset shows averaged spectra from the lower (red) and upper (black) copper terrace, demonstrating the minute difference of peak energies determined by the fitting procedure.

Also, the peaks on the Co islands are shifted towards higher energy, if at all, so exactly opposite to the expectation. It has previously been demonstrated how different photon wavelengths are excited with different probability due to variations in the inelastic channel. The fact that the Friedel oscillations are visible on the map of the fitted peak positions demonstrates how this effect also causes apparent peak shifts in the order of 1 to 2 meV. Therefore, it comes at no surprise that also differences of the LDOS across the different materials cause an apparent peak shift. As long as the influence of the cavity resonance is similar or smaller compared to the influence of the energy-dependent excitation probability on the observed peak shift, one would have to know the precise energy-dependence of the inelastic transition probabilities in order to disentangle those two effects, which is a hopeless endeavour. Alternatively, much more significant shifts would have to be observed in order to unambiguously ascribe them to the variation of the electromagnetic cavity response, which apparently requires a heteroepitaxial system with thicker islands. It can be concluded, that there is no significant variation of the cavity properties among the Cu area and the Co islands, so that this effect cannot significantly contribute to the observed intensity contrast.

Finally, the peak centre energies extracted from the Cu area aggregate to two distinct peaks in Fig. 5.17, instead of one. It was found that those peaks correspond to spectra from the lower and upper Cu terrace in the picture. To demonstrate that this is not an artefact from the fitting procedure, all emission spectra were averaged for both Cu terraces separately and compared in the inset of Fig. 5.17. The averaged spectrum from the lower terrace indeed extends further to lower wavelengths and is also slightly brighter, whereas the right sides of the spectra perfectly align. This, for once, demonstrates how well the fitting procedure works which was performed on the individual noisy curves after all. Physically, however, different parts of the Cu substrate should be equivalent and yield the same light emission, considering that the distance of the probed areas from any cavity altering surface features such as step edges was considerably larger than the plasmon confinement length. Due to the order, in which the individual points of the photon map were recorded (column-wise from left to right), one might suspect that the tip has slightly changed during the measurement. However, tip changes are easily recognisable in the maps, and no discontinuity other than the one shown in Fig. 5.12(b) could be found. Therefore, either a *series* of tip changes occurred, each of which too subtle to be identified individually, or different plasmon modes which are hosted by the cavity provide different spatial resolutions, resulting in minute variations of the cavity caused by surface features further away than the main resonances confinement. However, this issue cannot be resolved with the available data.

In summary, the only discussed effect which is capable of explaining the observed intensity contrast is the variation of the inelastic tunnelling probability. No evidence could be found that the material variation in the tunnelling junction had any measurable effect on the cavity resonance. This is surprising, considering that even with an assumed skin depth of 10 nm, the cobalt content of the involved material in the sample was 4 and 6 % when the tip was placed above the different islands, respectively, and the charge carrier density on the surface demonstrably varied across the substrate and the islands. However, a large portion of the charge carrier oscillation occurs at the tip, which was not altered during the experiment, which further reduces the contribution of the nano-islands to the total optically active material.

6 Discussion and conclusion

This thesis presented an innovative approach to collect light that is emitted from the tunnelling junction of an STM, which involves an integrated mirror tip consisting of an STM tip and a fully surrounding parabolic mirror, fabricated as one miniaturised part using DLW. Thereby, the requirement of repeatedly aligning and focussing the collection optics *in-situ* is eliminated, resulting in a greatly reduced number and complexity of additional components inside the STM in comparison to other STML setups. This limits possible sources of error and improves the usability. The combined mirror tip is small enough to couple the collected light directly into an optical fibre, thus, allowing for a simple light transmission path to the detector and avoiding the disadvantages related to free-space routing. Additionally, the alignment of the tip and the collection optics cannot drift during the experiment because the light source and the collecting mirror are always at the same, constant temperature. Combined with the excellent long-term stability of the STM, this allows for unprecedented long-term measurements during which the experimental conditions remain unchanged. This was demonstrated by acquiring a photon map over the course of ~30 h with minimal tip drift and a highly reproducible light signal, which is ten times longer than any other long-term measurement reported in the literature.

The mirror tips are employed in a custom STM which was specifically designed and assembled for this purpose and allows simple *in-situ* tip and sample transfers that are done within minutes. The alignment of the mirror tip relative to the fibre is established during tip changes by a modified optical ferrule installed in the scan head of the STM and a fitting, modified mating sleeve on the tip holder. For handling in the UHV chamber, the tip holders are placed on magnetic transfer plates from which they are simply picked up by the scan head, whereas the mating sleeve ensures axial and vertical alignment without any further adjustments. The STM was integrated into a newly assembled UHV machine with a JT style cryostat, which allows high sample-throughput due to low cool-down times as well as continuous measurements at 1.3 K for up to 4.5 days. This is the first STML setup which operates below 4 K and offers variable temperature operation, which will be beneficial for molecular studies in the future.

The newly conceived and assembled STM was characterised and optimised with regard to its long-term and short-term stability and proved very reliable, so far. For

STM operation, the mirror tips were found to be fully functional, albeit more difficult to work with compared to etched metal tips due to frequent changes of the atomic configurations at the tip apex and the limited lifetime of the metal coating. If required, improvements of the STM performance and light emission properties could be achieved by *in-situ* dipping into the surface or applying voltage pulses to some extent, and by reapplying the silver coat.

The mirror tips produced intense plasmon-mediated emission signals on Au(111) and Cu(111) surfaces which were comparable to or higher than most published figures. Still, the collection efficiency is not yet at its full potential and future optimisations might increase the achievable detection rates even further. Also the repeatability of the tip radius requires further optimisation as only few mirror tips featured the signature associated with narrow plasmon confinement, i.e., few resonances which are at high energy. Two voltage maps were recorded on Au(111) that showed the light emission depending on the bias voltage at constant current. They demonstrated the dominating influence of the tip shape on many aspects of the light emission due to the interplay of the cavity resonances and the tip-dependent voltage-distance characteristics. A spatially resolved photon map on Au(111) reproduced findings of the literature regarding reduced emission intensity at steps and on adsorbates. The expected shift of the cavity resonance at the step edge was only observed in one direction and set in at a large distance from the step edges, indicating a low and possibly asymmetric plasmon confinement. The tip configuration has changed several times throughout this measurement resulting in qualitatively different energy shift behaviour directly at the step edge. One type of tip configurations produced a smooth transition of the resonance energy across the step edge similar to reference measurements in the literature. Another type resulted in a sharply localised red-shift, which was never observed before and can be understood as a consequence of the energy-dependent excitation probability, which is different for different tip configurations.

The newly devised instrument was finally used to study the plasmon-mediated light emission from Co nano-islands grown on a Cu(111) surface. This is the first STML study of a heteroepitaxial system consisting of two metallic components, and light emission could be observed from the bare Cu substrate as well as from bilayer and trilayer Co islands side by side in a single experiment. This way, the material influence on the light emission could be studied under the same experimental conditions in a comparative fashion, in particular ruling out variations due to tip changes. A single high resolution photon map with high signal-to-noise ratio allowed to observe a variety of effects which were, so far, never observed at once. As concluded from the cavity alteration close to step edges, the employed mirror tip featured very high resolution not only for STM imaging but also in the sense that the optically active area was confined to a radius of 1.5 nm. This greatly demonstrates the usability of the covered polymer tips for STML. As the photon map was acquired across step edges of different heights, it was furthermore observed for the first time how this geometric cavity alteration amplified the associated resonance shift. Apart from these small resonance shifts close to the step edges, the overall shape of the emission spectra was found to be unaffected by the different materials. The emission intensity from the Co islands was, however, reduced by a characteristic, energy-dependent factor compared to the emission obtained from the Cu area. It was demonstrated by elastic spectroscopy that the probability of the corresponding tunnelling transitions is very tip dependent, but often follows a similar trend. Even though it was not possible to perform STS with the same tip configuration that was present during the STML measurement, it is plausible that the observed intensity contrast can be mostly attributed to variations of the excitation probability. A comparison with STS spectra from the literature and one spectrum after the tip has changed indicates that it might even be possible to find the quantitative ratios of the emission intensities reproduced in corresponding spectra of the differential conductance, which was never attempted before and might be a promising approach for future studies.

Other potential influences on the light emission, such as the material-dependent tip-sample separation and the different dielectric constants of Co and Cu, could be shown to be negligible in comparison. The latter was concluded from the lack of a significant energy shift of the gap plasmon if the tip was moved from the Cu substrate to the nano-islands. The minute shifts that have been observed can be well explained by the energy-dependent excitation probability, too. This means that three atomic layers of alien material were not sufficient to significantly tune the electromagnetic response of the gap cavity.

Additionally, it was observed that the emission intensity was modulated by Friedel oscillations of the surface states of the Cu area as well as the Co islands, which has been reported only once before on a Au(111) surface. Because different emitted photon wavelengths arise from inelastic transitions with different end states, isochromatic maps at different photon wavelengths featured electronic standing waves with different wavelengths as well. This allowed to quantitatively extract the dispersion of the surface states on the Cu area and a trilayer island in a wide energy range in one single measurement. The obtained dispersion relations agreed well with published dispersions obtained with STS, and the dispersion on the trilayer island exhibited a discrete spectrum due to the lateral confinement.

With the γ-STM, an entirely new and innovative experimental approach for collecting photons from the STM junction was demonstrated. Its operation principle has the potential to be the most efficient and most convenient design for this purpose once the beam collimation capability of the mirror and the stability of the tip coating are fully optimised. In the future, this new instrument will be used to investigate the light emission from self-decoupling molecules in order to understand the required chemistry for efficient electron-to-light conversion at the smallest scale, as well as the complex interplay of the dynamics of conduction electrons in the metallic leads and transitions between molecular states.

Bibliography

[1] G. Binnig, H. Rohrer, C. Gerber, and E. Weibel, "Tunneling through a controllable vacuum gap", *Appl. Phys. Lett.* **40**, 178–180 (1982).

[2] G. Binnig, H. Rohrer, C. Gerber, and E. Weibel, "Surface studies by scanning tunneling microscopy", *Phys. Rev. Lett.* **49**, 57–61 (1982).

[3] G. Binnig, H. Rohrer, C. Gerber, and E. Weibel, "7×7 reconstruction on Si(111) resolved in real space", *Phys. Rev. Lett.* **50**, 120–123 (1983).

[4] R. M. Feenstra, "Scanning tunneling spectroscopy", *Surf. Sci.* **299-300**, 965–979 (1994).

[5] D. P. E. Smith, G. Binnig, and C. F. Quate, "Detection of phonons with a scanning tunneling microscope", *Appl. Phys. Lett.* **49**, 1641–1643 (1986).

[6] D. P. E. Smith, M. D. Kirk, and C. F. Quate, "Molecular images and vibrational spectroscopy of sorbic acid with the scanning tunneling microscope", *J. Chem. Phys.* **86**, 6034–6038 (1987).

[7] C. Joachim, J. K. Gimzewski, R. R. Schlittler, and C. Chavy, "Electronic transparence of a single C_{60} molecule", *Phys. Rev. Lett.* **74**, 2102–2105 (1995).

[8] E. Ganz, K. Sattler, and J. Clarke, "Scanning tunneling microscopy of silver, gold, and aluminum monomers and small clusters on graphite", *J. Vac. Sci. Technol. A* **6**, 419–423 (1988).

[9] A. L. de Lozanne, S. A. Elrod, and C. F. Quate, "Spatial variations in the superconductivity of Nb_3Sn measured by low-temperature tunneling microscopy", *Phys. Rev. Lett.* **54**, 2433–2436 (1985).

[10] R. S. Becker, J. A. Golovchenko, and B. S. Swartzentruber, "Tunneling images of germanium surface reconstructions and phase boundaries", *Phys. Rev. Lett.* **54**, 2678–2680 (1985).

[11] D. P. Smith, A. Bryant, C. F. Quate, J. P. Rabe, C. Gerber, and J. D. Swalen, "Images of a lipid bilayer at molecular resolution by scanning tunneling microscopy.", *Proc. Natl. Acad. Sci. U.S.A.* **84**, 969–972 (1987).

[12] R. Wiesendanger, H.-J. Güntherodt, G. Güntherodt, R. J. Gambino, and R. Ruf, "Observation of vacuum tunneling of spin-polarized electrons with the scanning tunneling microscope", *Phys. Rev. Lett.* **65**, 247–250 (1990).

[13] D. M. Eigler and E. K. Schweizer, "Positioning single atoms with a scanning tunnelling microscope", *Nature* **344**, 524–526 (1990).

[14] L. Kuipers, R. W. M. Loos, H. Neerings, J. ter Horst, G. J. Ruwiel, A. P. de Jongh, and J. W. M. Frenken, "Design and performance of a high-temperature, high-speed scanning tunneling microscope", *Rev. Sci. Instrum.* **66**, 4557–4565 (1995).

[15] S. Müllegger, S. Tebi, A. K. Das, W. Schöfberger, F. Faschinger, and R. Koch, "Radio frequency scanning tunneling spectroscopy for single-molecule spin resonance", *Phys. Rev. Lett.* **113**, 133001 (2014).

[16] G. Nunes and M. R. Freeman, "Picosecond resolution in scanning tunneling microscopy", *Science* **262**, 1029–1032 (1993).

[17] S.-W. Hla and K.-H. Rieder, "STM control of chemical reactions: single-molecule synthesis", *Annu. Rev. Phys. Chem.* **54**, 307–330 (2003).

[18] M. Lackinger and W. M. Heckl, "A STM perspective on covalent intermolecular coupling reactions on surfaces", *J. Phys. D: Appl. Phys.* **44**, 464011 (2011).

[19] J. Lambe and S. L. McCarthy, "Light emission from inelastic electron tunneling", *Phys. Rev. Lett.* **37**, 923–925 (1976).

[20] J. H. Coombs, J. K. Gimzewski, B. Reihl, J. K. Sass, and R. R. Schlittler, "Photon emission experiments with the scanning tunnelling microscope", *J. Microsc.* **152**, 325–336 (1988).

[21] J. K. Gimzewski, B. Reihl, J. H. Coombs, and R. R. Schlittler, "Photon emission with the scanning tunneling microscope", *Z. Phys. B* **72**, 497–501 (1988).

[22] J. K. Gimzewski, J. K. Sass, R. R. Schlitter, and J. Schott, "Enhanced photon emission in scanning tunnelling microscopy", *EPL* **8**, 435–440 (1989).

[23] D. L. Abraham, A. Veider, C. Schönenberger, H. P. Meier, D. J. Arent, and S. F. Alvarado, "Nanometer resolution in luminescence microscopy of III-V heterostructures", *Appl. Phys. Lett.* **56**, 1564–1566 (1990).

[24] R. Berndt, A. Baratoff, and J. K. Gimzewski, "Scanning tunneling optical microscopy (STOM) of silver nanostructures", *Scanning Tunneling Microscopy and Related Methods*, Springer Netherlands, 1990, pp. 269–280.

[25] R. Berndt, R. R. Schlittler, and J. K. Gimzewski, "Photon emission scanning tunneling microscope", *J. Vac. Sci. Technol. B* **9**, 573 (1991).

[26] R. Berndt, J. K. Gimzewski, and P. Johansson, “Inelastic tunneling excitation of tip-induced plasmon modes on noble-metal surfaces”, *Phys. Rev. Lett.* **67**, 3796–3799 (1991).

[27] J. K. Gimzewski, R. Berndt, and R. R. Schlittler, “Observation of local photoemission using a scanning tunneling microscope”, *Ultramicroscopy* **42-44**, 366–370 (1992).

[28] R. Berndt and J. K. Gimzewski, “The role of proximity plasmon modes on noble metal surfaces in scanning tunneling microscopy”, *Surf. Sci.* **269-270**, 556–559 (1992).

[29] R. Berndt and J. K. Gimzewski, “Photon emission in scanning tunneling microscopy: interpretation of photon maps of metallic systems”, *Phys. Rev. B* **48**, 4746–4754 (1993).

[30] R. Berndt, J. K. Gimzewski, and P. Johansson, “Electromagnetic interactions of metallic objects in nanometer proximity”, *Phys. Rev. Lett.* **71**, 3493–3496 (1993).

[31] C. Zhang, L. Chen, R. Zhang, and Z. Dong, “Scanning tunneling microscope based nanoscale optical imaging of molecules on surfaces”, *Jpn. J. Appl. Phys.* **54**, 08LA01 (2015).

[32] K. Kuhnke, C. Große, P. Merino, and K. Kern, “Atomic-scale imaging and spectroscopy of electroluminescence at molecular interfaces”, *Chem. Rev.* **117**, 5174 (2017).

[33] S. F. Alvarado, “Luminescence in scanning tunneling microscopy on III–V nanostructures”, *J. Vac. Sci. Technol. B* **9**, 409 (1991).

[34] M. Reinhardt, G. Schull, P. Ebert, and R. Berndt, “Atomic resolution in tunneling induced light emission from GaAs(110)”, *Appl. Phys. Lett.* **96**, 152107 (2010).

[35] T. Lutz, A. Kabakchiev, T. Dufaux, C. Wolpert, Z. Wang, M. Burghard, K. Kuhnke, and K. Kern, “Scanning tunneling luminescence of individual CdSe nanowires”, *Small* **7**, 2396–2400 (2011).

[36] R. Coratger, J.-P. Salvetat, A. Carladous, F. Ajustron, J. Beauvillain, J.-M. Bonard, and L. Forró, “STM induced luminescence in carbon nanotubes”, *Eur. Phys. J. Appl. Phys.* **15**, 177–180 (2001).

[37] T. Uemura, S. Yamaguchi, M. Akai-Kasaya, A. Saito, M. Aono, and Y. Kuwahara, “Tunneling-current-induced light emission from individual carbon nanotubes”, *Surf. Sci.* **600**, L15–L19 (2006).

[38] S. Katano, H. Fujita, and Y. Uehara, "Investigation of local modification and luminescence of a carbon nanotube by scanning tunneling microscopy", *Appl. Phys. Lett.* **112**, 011601 (2018).

[39] P. Merino, C. Große, A. Rosławska, K. Kuhnke, and K. Kern, "Exciton dynamics of C_{60}-based single-photon emitters explored by Hanbury Brown-Twiss scanning tunnelling microscopy", *Nat. Commun.* **6**, 8461 (2015).

[40] C. Große, O. Gunnarsson, P. Merino, K. Kuhnke, and K. Kern, "Nanoscale imaging of charge carrier and exciton trapping at structural defects in organic semiconductors", *Nano Lett.* **16**, 2084–2089 (2016).

[41] C. Große, P. Merino, A. Rosławska, O. Gunnarsson, K. Kuhnke, and K. Kern, "Submolecular electroluminescence mapping of organic semiconductors", *ACS Nano* **11**, 1230–1237 (2017).

[42] A. Rosławska, P. Merino, C. Große, C. C. Leon, O. Gunnarsson, M. Etzkorn, K. Kuhnke, and K. Kern, "Single charge and exciton dynamics probed by molecular-scale-induced electroluminescence", *Nano Lett.* **18**, 4001–4007 (2018).

[43] K. Kuhnke, V. Turkowski, A. Kabakchiev, T. Lutz, T. S. Rahman, and K. Kern, "Pentacene excitons in strong electric fields", *ChemPhysChem* **19**, 277–283 (2018).

[44] P. Merino, A. Rosławska, C. Große, C. C. Leon, K. Kuhnke, and K. Kern, "Bimodal exciton-plasmon light sources controlled by local charge carrier injection", *Sci. Adv.* **4**, eaap8349 (2018).

[45] B. N. J. Persson and A. Baratoff, "Theory of photon emission in electron tunneling to metallic particles", *Phys. Rev. Lett.* **68**, 3224–3227 (1992).

[46] P. Johansson, "Light emission from a scanning tunneling microscope: fully retarded calculation", *Phys. Rev. B* **58**, 10823–10834 (1998).

[47] G. Hoffmann, J. Kröger, and R. Berndt, "Color imaging with a low temperature scanning tunneling microscope", *Rev. Sci. Instrum.* **73**, 305–309 (2002).

[48] D. Gramotnev and S. Bozhevolnyi, "Plasmonics beyond the diffraction limit", *Nat. Photonics* **4**, 83–91 (2010).

[49] R. Esteban, R. Vogelgesang, J. Dorfmüller, A. Dmitriev, C. Rockstuhl, C. Etrich, and K. Kern, "Direct near-field optical imaging of higher order plasmonic resonances", *Nano Lett.* **8**, 3155–3159 (2008).

[50] F. J. G. de Abajo, "Optical excitations in electron microscopy", *Rev. Mod. Phys.* **82**, 209–275 (2010).

[51] L. Douillard and F. Charra, “High-resolution mapping of plasmonic modes: photoemission and scanning tunnelling luminescence microscopies”, *J. Phys. D: Appl. Phys.* **44**, 464002 (2011).

[52] G. Hoffmann and R. Berndt, “Electromagnetic coupling at an atomic scale”, *Appl. Phys. A* **72**, S173–S175 (2001).

[53] Y. Uehara and S. Ushioda, “Atomic-site-dependent light emission from Au (110)-(2×1) surface induced by scanning tunneling microscope”, *Phys. Rev. B* **66**, 165420 (2002).

[54] J. Aizpurua, G. Hoffmann, S. P. Apell, and R. Berndt, “Electromagnetic coupling on an atomic scale”, *Phys. Rev. Lett.* **89**, 156803 (2002).

[55] G. Hoffmann, T. Maroutian, and R. Berndt, “Color view of atomic highs and lows in tunneling induced light emission”, *Phys. Rev. Lett.* **93**, 076102 (2004).

[56] M. G. Boyle, J. Mitra, and P. Dawson, “Photon emission at step edges of single crystalline gold surfaces investigated by scanning tunnelling microscopy”, *Jpn. J. Appl. Phys.* **45**, 2119–2123 (2006).

[57] R. Berndt, R. Gaisch, W. D. Schneider, J. K. Gimzewski, B. Reihl, R. R. Schlittler, and M. Tschudy, “Atomic resolution in photon emission induced by a scanning tunneling microscope”, *Phys. Rev. Lett.* **74**, 102–105 (1995).

[58] L. Pascua, F. Stavale, N. Nilius, and H.-J. Freund, “Ag/ZnO hybrid systems studied with scanning tunnelling microscopy-based luminescence spectroscopy”, *J. Appl. Phys.* **119**, 095310 (2016).

[59] Y.-H. Liau and N. F. Scherer, “Mechanism for photon emission from Au nanohemispheres induced by scanning tunneling microscopy”, *Appl. Phys. Lett.* **74**, 3966–3968 (1999).

[60] P. Dumas, C. Syrykh, I. V. Makarenko, and F. Salvan, “STM-induced light emission of supported silver nanocrystallites”, *EPL* **40**, 447–452 (1997).

[61] S. Divitt, P. Bharadwaj, and L. Novotny, “The role of gap plasmons in light emission from tunnel junctions”, *Opt. Express* **21**, 27452 (2013).

[62] A. Downes, P. Guaino, and P. Dumas, “Color mapping by scanning tunneling microscopy: chemical analysis of metal surfaces”, *Appl. Phys. Lett.* **80**, 380–382 (2002).

[63] A. Yu, S. Li, G. Czap, and W. Ho, “Tunneling-electron-induced light emission from single gold nanoclusters”, *Nano Lett.* **16**, 5433–5436 (2016).

[64] E. L. Moal, S. Marguet, B. Rogez, S. Mukherjee, P. D. Santos, E. Boer-Duchemin, G. Comtet, and G. Dujardin, “An electrically excited nanoscale light source with active angular control of the emitted light”, *Nano Lett.* **13**, 4198–4205 (2013).

[65] C. Maurel, R. Coratger, F. Ajustron, G. Seine, R. Péchou, and J. Beauvillain, “Photon emission from STM of granular gold in UHV: comparison with air and study of spectra shifting with tip position”, *Eur. Phys. J. Appl. Phys.* **21**, 121–126 (2002).

[66] P. Myrach, N. Nilius, and H.-J. Freund, “Photon mapping of individual Ag particles on MgO/Mo(001)”, *Phys. Rev. B* **83**, 035416 (2011).

[67] N. Nilius, N. Ernst, and H.-J. Freund, “Photon emission spectroscopy of individual oxide-supported silver clusters in a scanning tunneling microscope”, *Phys. Rev. Lett.* **84**, 3994–3997 (2000).

[68] N. Nilius, N. Ernst, and H.-J. Freund, “Tip influence on plasmon excitations in single gold particles in an STM”, *Phys. Rev. B* **65**, 115421 (2002).

[69] Z. Dong, H.-S. Chu, D. Zhu, W. Du, Y. A. Akimov, W. P. Goh, T. Wang, K. E. J. Goh, C. Troadec, C. A. Nijhuis, and J. K. W. Yang, “Electrically-excited surface plasmon polaritons with directionality control”, *ACS Photonics* **2**, 385–391 (2015).

[70] U. D. Keil, T. Ha, J. R. Jensen, and J. M. Hvam, “Femtosecond tunneling response of surface plasmon polaritons”, *Appl. Phys. Lett.* **72**, 3074–3076 (1998).

[71] C. Große, M. Etzkorn, K. Kuhnke, S. Loth, and K. Kern, “Quantitative mapping of fast voltage pulses in tunnel junctions by plasmonic luminescence”, *Appl. Phys. Lett.* **103**, 183108 (2013).

[72] D. Waldeck, A. P. Alivisatos, and C. B. Harris, “Nonradiative damping of molecular electronic excited states by metal surfaces”, *Surf. Sci.* **158**, 103–125 (1985).

[73] W. L. Barnes, “Fluorescence near interfaces: the role of photonic mode density”, *J. Mod. Opt.* **45**, 661–699 (1998).

[74] S. Ho, S. Liu, Y. Chen, and F. So, “Review of recent progress in multilayer solution-processed organic light-emitting diodes”, *J. Photonics Energy* **5**, 057611 (2015).

[75] E. Ćavar, M.-C. Blüm, M. Pivetta, F. Patthey, M. Chergui, and W.-D. Schneider, “Fluorescence and phosphorescence from individual C_{60} molecules excited by local electron tunneling”, *Phys. Rev. Lett.* **95**, 196102 (2005).

[76] J. Lee, S. M. Perdue, A. R. Perez, and V. A. Apkarian, “Vibronic motion with joint angstrom-femtosecond resolution observed through fano progressions recorded within one molecule”, *ACS Nano* **8**, 54–63 (2014).

[77] B. Doppagne, M. C. Chong, E. Lorchat, S. Berciaud, M. Romeo, H. Bulou, A. Boeglin, F. Scheurer, and G. Schull, "Vibronic spectroscopy with submolecular resolution from STM-induced electroluminescence", *Phys. Rev. Lett.* **118**, 127401 (2017).

[78] B. Doppagne, M. C. Chong, H. Bulou, A. Boeglin, F. Scheurer, and G. Schull, "Electrofluorochromism at the single molecule level", *Science* **361**, 251–255 (2018).

[79] A. Yu, S. Li, B. Dhital, H. P. Lu, and W. Ho, "Tunneling electron induced charging and light emission of single panhematin molecules", *J. Phys. Chem. C* **120**, 21099–21103 (2016).

[80] M. C. Chong, L. Sosa-Vargas, H. Bulou, A. Boeglin, F. Scheurer, F. Mathevet, and G. Schull, "Ordinary and hot electroluminescence from single-molecule devices: controlling the emission color by chemical engineering", *Nano Lett.* **16**, 6480–6484 (2016).

[81] H. Imada, K. Miwa, M. Imai-Imada, S. Kawahara, K. Kimura, and Y. Kim, "Real-space investigation of energy transfer in heterogeneous molecular dimers", *Nature* **538**, 364–367 (2016).

[82] Y. Zhang, Y. Luo, Y. Zhang, Y.-J. Yu, Y.-M. Kuang, L. Zhang, Q.-S. Meng, Y. Luo, J.-L. Yang, Z.-C. Dong, and J. G. Hou, "Visualizing coherent intermolecular dipole-dipole coupling in real space", *Nature* **531**, 623–627 (2016).

[83] M. C. Chong, G. Reecht, H. Bulou, A. Boeglin, F. Scheurer, F. Mathevet, and G. Schull, "Narrow-line single-molecule transducer between electronic circuits and surface plasmons", *Phys. Rev. Lett.* **116**, 036802 (2016).

[84] H. Imada, K. Miwa, M. Imai-Imada, S. Kawahara, K. Kimura, and Y. Kim, "Single-molecule investigation of energy dynamics in a coupled plasmon-exciton system", *Phys. Rev. Lett.* **119**, 013901 (2017).

[85] A. Yu, S. Li, H. Wang, S. Chen, R. Wu, and W. Ho, "Visualization of nanoplasmonic coupling to molecular orbital in light emission induced by tunneling electrons", *Nano Lett.* **18**, 3076–3080 (2018).

[86] K. Kuhnke, A. Kabakchiev, W. Stiepany, F. Zinser, R. Vogelgesang, and K. Kern, "Versatile optical access to the tunnel gap in a low-temperature scanning tunneling microscope", *Rev. Sci. Instrum.* **81**, 113102 (2010).

[87] Y. Khang, Y. Park, M. Salmeron, and E. R. Weber, "Low temperature ultrahigh vacuum cross-sectional scanning tunneling microscope for luminescence measurements", *Rev. Sci. Instrum.* **70**, 4595–4599 (1999).

[88] S. Kawata, H.-B. Sun, T. Tanaka, and K. Takada, "Finer features for functional microdevices", *Nature* **412**, 697–698 (2001).

[89] J. Bardeen, "Tunnelling from a many-particle point of view", *Phys. Rev. Lett.* **6**, 57–59 (1961).

[90] J. Tersoff and D. R. Hamann, "Theory and application for the scanning tunneling microscope", *Phys. Rev. Lett.* **50**, 1998–2001 (1983).

[91] T. E. Feuchtwang, P. H. Cutler, and N. M. Miskovsky, "A theory of vacuum tunneling microscopy", *Phys. Lett. A* **99**, 167–171 (1983).

[92] A. Selloni, P. Carnevali, E. Tosatti, and C. D. Chen, "Voltage-dependent scanning-tunneling microscopy of a crystal surface: graphite", *Scanning Tunneling Microscopy*, Springer Netherlands, 1985, pp. 168–171.

[93] N. D. Lang, "Spectroscopy of single atoms in the scanning tunneling microscope", *Phys. Rev. B* **34**, 5947–5950 (1986).

[94] V. A. Ukraintsev, "Data evaluation technique for electron-tunneling spectroscopy", *Phys. Rev. B* **53**, 11176–11185 (1996).

[95] A. Heinrich, "Single-atom spin-flip spectroscopy", *Science* **306**, 466–469 (2004).

[96] J. Fransson, "Spin inelastic electron tunneling spectroscopy on local spin adsorbed on surface", *Nano Lett.* **9**, 2414–2417 (2009).

[97] T. Balashov, A. F. Takács, W. Wulfhekel, and J. Kirschner, "Magnon excitation with spin-polarized scanning tunneling microscopy", *Phys. Rev. Lett.* **97**, 187201 (2006).

[98] P. J. Bryant, H. S. Kim, Y. C. Zheng, and R. Yang, "Technique for shaping scanning tunneling microscope tips", *Rev. Sci. Instrum.* **58**, 1115–1115 (1987).

[99] G. W. Stupian and M. S. Leung, "A scanning tunneling microscope based on a motorized micrometer", *Rev. Sci. Instrum.* **60**, 181–185 (1989).

[100] G. Binnig and D. P. E. Smith, "Single-tube three-dimensional scanner for scanning tunneling microscopy", *Rev. Sci. Instrum.* **57**, 1688–1689 (1986).

[101] S. H. Pan, E. W. Hudson, and J. C. Davis, "^{3}He refrigerator based very low temperature scanning tunneling microscope", *Rev. Sci. Instrum.* **70**, 1459–1463 (1999).

[102] C. Renner, P. Niedermann, A. D. Kent, and Ø. Fischer, "A vertical piezoelectric inertial slider", *Rev. Sci. Instrum.* **61**, 965–967 (1990).

[103] L. Zhang, T. Miyamachi, T. Tomanić, R. Dehm, and W. Wulfhekel, "A compact sub-Kelvin ultrahigh vacuum scanning tunneling microscope with high energy resolution and high stability", *Rev. Sci. Instrum.* **82**, 103702 (2011).

[104] B. Koslowski, C. Dietrich, A. Tschetschetkin, and P. Ziemann, "Design of an extremely stable low-temperature ultrahigh vacuum scanning tunneling microscope", *Rev. Sci. Instrum.* **77**, 063707 (2006).

[105] K. Besocke, "An easily operable scanning tunneling microscope", *Surf. Sci.* **181**, 145–153 (1987).

[106] N. W. Ashcroft and D. N. Mermin, "Festkörperphysik", Third Ed., Oldenbourg Wissenschaftsverlag, 2007, ISBN: 978-3-486-58273-4.

[107] H. Raether, "Surface Plasmons on Smooth and Rough Surfaces and on Gratings", Springer Berlin Heidelberg, 1988, ISBN: 978-3-540-17363-2.

[108] L. Novotny and B. Hecht, "Principles of Nano-Optics", Cambridge University Press, 2006, ISBN: 0521832241.

[109] E. Kretschmann and H. Raether, "Notizen: radiative decay of non radiative surface plasmons excited by light", *Z. Naturforsch. A* **23**, 2135 (1968).

[110] F. Ding, Y. Yang, R. A. Deshpande, and S. I. Bozhevolnyi, "A review of gap-surface plasmon metasurfaces: fundamentals and applications", *Nanophotonics* **7**, 1129 (2018).

[111] R. W. Rendell and D. J. Scalapino, "Surface plasmons confined by microstructures on tunnel junctions", *Phys. Rev. B* **24**, 3276–3294 (1981).

[112] G. Schull, N. Néel, P. Johansson, and R. Berndt, "Electron-plasmon and electron-electron interactions at a single atom contact", *Phys. Rev. Lett.* **102**, 057401 (2009).

[113] N. L. Schneider, G. Schull, and R. Berndt, "Optical probe of quantum shot-noise reduction at a single-atom contact", *Phys. Rev. Lett.* **105**, 026601 (2010).

[114] K. Meguro, K. Sakamoto, R. Arafune, M. Satoh, and S. Ushioda, "Origin of multiple peaks in the light emission spectra of a Au(111) surface induced by the scanning tunneling microscope", *Phys. Rev. B* **65**, 165405 (2002).

[115] M. G. Boyle, J. Mitra, and P. Dawson, "Infrared emission from tunneling electrons: the end of the rainbow in scanning tunneling microscopy", *Appl. Phys. Lett.* **94**, 233118 (2009).

[116] G. Hoffmann, J. Aizpurua, P. Apell, and R. Berndt, "Influence of tip geometry in light emission from the scanning tunnelling microscope", *Surf. Sci.* **482-485**, 1159–1162 (2001).

[117] R. Berndt and J. K. Gimzewski, "Isochromat spectroscopy of photons emitted from metal surfaces in an STM", *Ann. Phys.* **505**, 133–140 (1993).

[118] D. Hone, B. Mühlschlegel, and D. J. Scalapino, "Theory of light emission from small particle tunnel junctions", *Appl. Phys. Lett.* **33**, 203–204 (1978).

[119] V. Kalathingal, P. Dawson, and J. Mitra, "Scanning tunnelling microscope light emission: finite temperature current noise and over cut-off emission", *Sci. Rep.* **7**, 3530 (2017).

[120] F. Xu, C. Holmqvist, and W. Belzig, "Overbias light emission due to higher-order quantum noise in a tunnel junction", *Phys. Rev. Lett.* **113**, 066801 (2014).

[121] G. Schull, M. Becker, and R. Berndt, "Imaging confined electrons with plasmonic light", *Phys. Rev. Lett.* **101**, 136801 (2008).

[122] P. Chen, W. Wang, N. Lin, and S. Du, "Manipulating photon emission efficiency with local electronic states in a tunneling gap", *Opt. Express* **22**, 8234 (2014).

[123] D. Steinmüller-Nethl, R. A. Höpfel, E. Gornik, A. Leitner, and F. R. Aussenegg, "Femtosecond relaxation of localized plasma excitations in Ag islands", *Phys. Rev. Lett.* **68**, 389–392 (1992).

[124] T. Wang, E. Boer-Duchemin, Y. Zhang, G. Comtet, and G. Dujardin, "Excitation of propagating surface plasmons with a scanning tunnelling microscope", *Nanotechnology* **22**, 175201 (2011).

[125] P. Bharadwaj, A. Bouhelier, and L. Novotny, "Electrical excitation of surface plasmons", *Phys. Rev. Lett.* **106**, 226802 (2011).

[126] Y. Lu, Y. Chen, J. Xu, T. Wang, and J.-T. Lü, "Energy transfer channels of the plasmon excitation process in STM tunnel junctions", arXiv: `1710.02450v1 [physics.optics]` (2017).

[127] A. Hartschuh, "Tip-enhanced near-field optical microscopy", *Angew. Chem. Int. Ed.* **47**, 8178–8191 (2008).

[128] P. Johansson, R. Monreal, and P. Apell, "Theory for light emission from a scanning tunneling microscope", *Phys. Rev. B* **42**, 9210–9213 (1990).

[129] A. Madrazo, M. Nieto-Vesperinas, and N. García, "Exact calculation of Maxwell equations for a tip-metallic interface configuration: application to atomic resolution by photon emission", *Phys. Rev. B* **53**, 3654–3657 (1996).

[130] J. Aizpurua, S. P. Apell, and R. Berndt, "Role of tip shape in light emission from the scanning tunneling microscope", *Phys. Rev. B* **62**, 2065–2073 (2000).

[131] N. Behr and M. B. Raschke, "Optical antenna properties of scanning probe tips: plasmonic light scattering, tip-sample coupling, and near-field enhancement", *J. Phys. Chem. C* **112**, 3766–3773 (2008).

[132] L. G. Chen, C. Zhang, R. Zhang, X. L. Zhang, and Z. C. Dong, "Note: optical optimization for ultrasensitive photon mapping with submolecular resolution by scanning tunneling microscope induced luminescence", *Rev. Sci. Instrum.* **84**, 066106 (2013).

[133] P. Schmidt, R. Berndt, and M. I. Vexler, "Ultraviolet light emission from Si in a scanning tunneling microscope", *Phys. Rev. Lett.* **99**, 246103 (2007).

[134] G. Hoffmann, L. Libioulle, and R. Berndt, "Tunneling-induced luminescence from adsorbed organic molecules with submolecular lateral resolution", *Phys. Rev. B* **65**, 212107 (2002).

[135] X. Tao, Z. C. Dong, J. L. Yang, Y. Luo, J. G. Hou, and J. Aizpurua, "Influence of a dielectric layer on photon emission induced by a scanning tunneling microscope", *J. Chem. Phys.* **130**, 084706 (2009).

[136] C. Große, A. Kabakchiev, T. Lutz, R. Froidevaux, F. Schramm, M. Ruben, M. Etzkorn, U. Schlickum, K. Kuhnke, and K. Kern, "Dynamic control of plasmon generation by an individual quantum system", *Nano Lett.* **14**, 5693–5697 (2014).

[137] X. H. Qiu, G. V. Nazin, and W. Ho, "Vibrationally resolved fluorescence excited with submolecular precision", *Science* **299**, 542–546 (2003).

[138] N. L. Schneider and R. Berndt, "Plasmonic excitation of light emission and absorption by porphyrine molecules in a scanning tunneling microscope", *Phys. Rev. B* **86**, 035445 (2012).

[139] Z.-C. Dong, X.-L. Guo, A. S. Trifonov, P. S. Dorozhkin, K. Miki, K. Kimura, S. Yokoyama, and S. Mashiko, "Vibrationally resolved fluorescence from organic molecules near metal surfaces in a scanning tunneling microscope", *Phys. Rev. Lett.* **92**, 086801 (2004).

[140] E. Flaxer, O. Sneh, and O. Cheshnovsky, "Molecular light emission induced by inelastic electron tunneling", *Science* **262**, 2012–2014 (1993).

[141] H. Yang, A. J. Mayne, G. Comtet, G. Dujardin, Y. Kuk, S. Nagarajan, and A. Gourdon, "Single-molecule light emission at room temperature on a wide-band-gap semiconductor", *Phys. Rev. B* **90**, 125427 (2014).

[142] G. Reecht, F. Scheurer, V. Speisser, Y. J. Dappe, F. Mathevet, and G. Schull, "Electroluminescence of a polythiophene molecular wire suspended between a metallic surface and the tip of a scanning tunneling microscope", *Phys. Rev. Lett.* **112**, 047403 (2014).

[143] Z. Zhang, S. Sheng, R. Wang, and M. Sun, "Tip-enhanced Raman spectroscopy", *Anal. Chem.* **88**, 9328–9346 (2016).

[144] T. Murashita and M. Tanimoto, "Observation of tunneling electron luminescence at low temperatures using novel conductive transparent tip", *Jpn. J. Appl. Phys.* **34**, 4398–4400 (1995).

[145] J. G. Keizer, J. K. Garleff, and P. M. Koenraad, "Simple and efficient scanning tunneling luminescence detection at low-temperature", *Rev. Sci. Instrum.* **80**, 123704 (2009).

[146] N. Venkateswaran, K. Sattler, J. Xhie, and M. Ge, "Photon emission from nanogranular gold excited by electron tunneling", *Surf. Sci.* **274**, 199–204 (1992).

[147] V. Sivel, R. Coratger, F. Ajustron, and J. Beauvillain, "Photon emission stimulated by scanning tunneling microscopy in air", *Phys. Rev. B* **45**, 8634–8637 (1992).

[148] M. J. Gallagher, S. Howells, L. Yi, T. Chen, and D. Sarid, "Photon emission from gold surfaces in air using scanning tunneling microscopy", *Surf. Sci.* **278**, 270–280 (1992).

[149] S. Egusa, Y.-H. Liau, and N. F. Scherer, "Imaging scanning tunneling microscope-induced electroluminescence in plasmonic corrals", *Appl. Phys. Lett.* **84**, 1257–1259 (2004).

[150] S. Ushioda, "STM light emission spectroscopy of Si surfaces", *Solid State Commun.* **84**, 173–176 (1992).

[151] Z.-C. Dong, A. Kar, Z.-Q. Zou, T. Ohgi, P. Dorozhkin, D. Fujita, S. Yokoyama, T. Terui, T. Yamada, T. Kamikado, M. Zhou, S. Mashiko, and T. Okamoto, "Light emission from porphyrin molecules induced by a scanning tunneling microscope", *Jpn. J. Appl. Phys.* **41**, 4898–4902 (2002).

[152] M. J. Romero, J. van de Lagemaat, I. Mora-Sero, G. Rumbles, and M. M. Al-Jassim, "Imaging of resonant quenching of surface plasmons by quantum dots", *Nano Lett.* **6**, 2833–2837 (2006).

[153] Y. Suzuki, H. Minoda, and N. Yamamoto, "STM light emission from Ag/ Si(111)", *Surf. Sci.* **438**, 297–304 (1999).

[154] C. Thirstrup, M. Sakurai, and M. Aono, "Photon emission STM using optical fiber bunches", *J. Surf. Anal.* **44**, 152–158 (1998).

[155] M. Kemerink, J. W. Gerritsen, J. G. H. Hermsen, P. M. Koenraad, H. van Kempen, and J. Wolter, "Low-temperature scanning-tunneling microscope for luminescence measurements in high magnetic fields", *Rev. Sci. Instrum.* **72**, 132–135 (2001).

[156] R. Arafune, K. Sakamoto, K. Meguro, M. Satoh, A. Arai, and S. Ushioda, "Multiple-fiber collection system for scanning tunneling microscope light emission spectroscopy", *Jpn. J. Appl. Phys.* **40**, 5450–5453 (2001).

[157] N. J. Watkins, J. P. Long, Z. H. Kafafi, and A. J. Mäkinen, "Fiber optic light collection system for scanning-tunneling-microscope-induced light emission", *Rev. Sci. Instrum.* **78**, 053707 (2007).

[158] M. Sakurai, C. Thirstrup, and M. Aono, "Optical selection rules in light emission from the scanning tunneling microscope", *Phys. Rev. Lett.* **93**, 046102 (2004).

[159] S. Maruo, O. Nakamura, and S. Kawata, "Three-dimensional microfabrication with two-photon-absorbed photopolymerization", *Opt. Lett.* **22**, 132 (1997).

[160] T. Baldacchini, "Three-Dimensional Microfabrication Using Two-Photon Polymerization", William Andrew Publishing, 2015, 512 pp., ISBN: 0323353215.

[161] M. Thiel, M. S. Rill, G. von Freymann, and M. Wegener, "Three-dimensional bichiral photonic crystals", *Adv. Mater.* **21**, 4680–4682 (2009).

[162] J. K. Gansel, M. Thiel, M. S. Rill, M. Decker, K. Bade, V. Saile, G. von Freymann, S. Linden, and M. Wegener, "Gold helix photonic metamaterial as broadband circular polarizer", *Science* **325**, 1513–1515 (2009).

[163] T. Ergin, N. Stenger, P. Brenner, J. B. Pendry, and M. Wegener, "Three-dimensional invisibility cloak at optical wavelengths", *Science* **328**, 337–339 (2010).

[164] J. Mu, Z. Liu, J. Li, T. Hao, Y. Wang, S. Sun, Z.-Y. Li, J. Li, W. Li, and C. Gu, "Direct laser writing of pyramidal plasmonic structures with apertures and asymmetric gratings towards efficient subwavelength light focusing", *Opt. Express* **23**, 22564 (2015).

[165] J. H. Atwater, P. Spinelli, E. Kosten, J. Parsons, C. V. Lare, J. V. de Groep, J. G. de Abajo, A. Polman, and H. A. Atwater, "Microphotonic parabolic light directors fabricated by two-photon lithography", *Appl. Phys. Lett.* **99**, 151113 (2011).

[166] W. Yan, M. M. Hossain, and M. Gu, "High light-directing micrometer-sized parabolic mirror arrays", *Opt. Lett.* **38**, 3177 (2013).

[167] S. Juodkazis, "3D printed micro-optics", *Nat. Photonics* **10**, 499–501 (2016).

[168] F. Niesler and Y. Tanguy, "3D printers for the fabrication of micro-optical elements", *Optik & Photonik* **11**, 44–47 (2016).

[169] S. D. Kevan and W. Eberhardt, "Angle-Resolved Photoemission", ed. by S. D. Kevan, Elsevier S&T, 1992.

[170] W. Shockley, "On the surface states associated with a periodic potential", *Phys. Rev.* **56**, 317–323 (1939).

[171] P. M. Echenique and M. E. Uranga, "Image potential states at surfaces", *Surf. Sci.* **247**, 125–132 (1991).

[172] E. Bertel, "One- and two-dimensional surface states on metals", *Surf. Sci.* **331-333**, 1136–1146 (1995).

[173] M. F. Crommie, C. P. Lutz, and D. M. Eigler, "Imaging standing waves in a two-dimensional electron gas", *Nature* **363**, 524–527 (1993).

[174] Y. Hasegawa and P. Avouris, "Direct observation of standing wave formation at surface steps using scanning tunneling spectroscopy", *Phys. Rev. Lett.* **71**, 1071–1074 (1993).

[175] P. Avouris, "Real space imaging of electron scattering phenomena at metal surfaces", *J. Vac. Sci. Technol. B* **12**, 1447 (1994).

[176] K. Edelmann, L. Gerhard, M. Winkler, L. Wilmes, V. Rai, M. Schumann, C. Kern, M. Meyer, M. Wegener, and W. Wulfhekel, "Light collection from a low-temperature scanning tunneling microscope using integrated mirror tips fabricated by direct laser writing", *Rev. Sci. Instr.* **89**, 123107 (2018).

[177] M. Winkler, "Entwicklung eines optischen Rastertunnelmikroskops", MA thesis, Karlsruhe Institute of Technology (2016).

[178] L. Wilmes, "Messung der Oberflächenzustände von Co auf Cu(111) durch plasmonisches Licht unter Einsatz eines neuen Rastertunnelmikroskops mit optischem Zugang", Master's thesis, Karlsruhe Institute of Technology (2018).

[179] A. W. Schell, T. Neumer, Q. Shi, J. Kaschke, J. Fischer, M. Wegener, and O. Benson, "Laser-written parabolic micro-antennas for efficient photon collection", *Appl. Phys. Lett.* **105**, 231117 (2014).

[180] G. Göring, P.-I. Dietrich, M. Blaicher, S. Sharma, J. G. Korvink, T. Schimmel, C. Koos, and H. Hölscher, "Tailored probes for atomic force microscopy fabricated by two-photon polymerization", *Appl. Phys. Lett.* **109**, 063101 (2016).

[181] S. Hengsbach and A. D. Lantada, "Direct laser writing of auxetic structures: present capabilities and challenges", *Smart Mater. Struct.* **23**, 085033 (2014).

[182] T. Gissibl, S. Thiele, A. Herkommer, and H. Giessen, "Sub-micrometre accurate free-form optics by three-dimensional printing on single-mode fibres", *Nat. Commun.* **7**, 11763 (2016).

[183] G. Binnig and H. Rohrer, "Scanning tunneling microscopy", *Surf. Sci.* **126**, 236–244 (1983).

[184] C. Zhang, B. Gao, L. G. Chen, Q. S. Meng, H. Yang, R. Zhang, X. Tao, H. Y. Gao, Y. Liao, and Z. C. Dong, "Fabrication of silver tips for scanning tunneling microscope induced luminescence", *Rev. Sci. Instrum.* **82**, 083101 (2011).

[185] M. Schumann, T. Bückmann, N. Gruhler, M. Wegener, and W. Pernice, "Hybrid 2D–3D optical devices for integrated optics by direct laser writing", *Light Sci. Appl.* **3**, e175–e175 (2014).

[186] H. E. Bennett and J. O. Porteus, "Relation between surface roughness and specular reflectance at normal incidence", *J. Opt. Soc. Am.* **51**, 123 (1961).

[187] M. Czerny and A. F. Turner, "Über den Astigmatismus bei Spiegelspektrometern", *Z. Phys.* **61**, 792–797 (1930).

[188] N. J. Simon, E. S. Drexler, and R. P. Reed, "Properties of Copper and Copper Alloys at Cryogenic Temperatures", National Institute of Standards and Technology, 1992.

[189] P. E. Bradley and R. Radebaugh, "CRC handbook of chemistry and physics", ed. by W. M. Haynes, 94th ed., CRC Press, Boca Raton, Florida, 2013, chap. Properties of Selected Materials at Cryogenic Temperatures, ISBN: 9781466571150.

[190] W. R. G. Kemp, P. G. Klemens, and G. K. White, "Thermal and electrical conductivities of iron, nickel, titanium, and zirconium at low temperatures", *Aust. J. Phys.* **9**, 180 (1956).

[191] M. C. Runyan and W. C. Jones, "Thermal conductivity of thermally-isolating polymeric and composite structural support materials between 0.3 and 4 K", *Cryogenics* **48**, 448–454 (2008).

[192] Y. Kuk and P. J. Silverman, "Scanning tunneling microscope instrumentation", *Rev. Sci. Instrum.* **60**, 165–180 (1989).

[193] Manufacturer's specification by Little Falls Alloys, `http://www.lfa-wire.com/heat-treatable-alloy-25_c17200-and-c17300.htm`.

[194] S. A. Z. Jahromi, M. Salomons, Q. Sun, and R. A. Wolkow, "Prediction of the resonant frequency of piezoelectric tube scanners through three-dimensional finite element modeling of a tube assembly", *Rev. Sci. Instrum.* **79**, 076104 (2008).

[195] E. Mateo-Martí, C. Rogero, C. Gonzalez, J. M. Sobrado, P. L. de Andrés, and J. A. Martin-Gago, "Interplay between fast diffusion and molecular interaction in the formation of self-assembled nanostructures of S-cysteine on Au(111)", *Langmuir* **26**, 4113–4118 (2010).

[196] N. Battaglini, H. Klein, C. Hortholary, C. Coudret, F. Maurel, and P. Dumas, "STM observation of a single diarylethene flickering", *Ultramicroscopy* **107**, 958–962 (2007).

[197] K. Zenichowski, J. Dokić, T. Klamroth, and P. Saalfrank, "Current versus temperature-induced switching of a single molecule: open-system density matrix theory for 1,5-cyclooctadiene on Si(100)", *J. Chem. Phys.* **136**, 094705 (2012).

[198] K. Edelmann, "Untersuchung dreibeiniger Spirobifluoren-Moleküle auf einer Au(111)-Oberfläche mittels Rastertunnelmikroskopie", Diploma thesis, Karlsruhe Institute of Technology (2014).

[199] L. Gerhard, K. Edelmann, J. Homberg, M. Valášek, S. G. Bahoosh, M. Lukas, F. Pauly, M. Mayor, and W. Wulfhekel, "An electrically actuated molecular toggle switch", *Nat. Commun.* **8**, 14672 (2017).

[200] M. P. Edejer and G. Thodos, "Vapor pressures of liquid nitrogen between the triple and critical points", *J. Chem. Eng. Data* **12**, 206–209 (1967).

[201] H. T. Harmsel, H. V. Dijk, and M. Durieux, "The heat of vaporization of helium", *Physica* **36**, 620–636 (1967).

[202] R. Berman, "The thermal conductivity of some polycrystalline solids at low temperatures", *Proc. Phys. Soc. London, Sect. A* **65**, 1029–1040 (1952).

[203] A. Drobizhev, J. Reiten, V. Singh, and Y. G. Kolomensky, "Thermal conductivity measurements of PTFE and Al_2O_3 ceramic at sub-Kelvin temperatures", *Cryogenics* **85**, 63–70 (2016).

[204] T. Balashov, M. Meyer, and W. Wulfhekel, "A compact ultrahigh vacuum scanning tunneling microscope with dilution refrigeration", *Rev. Sci. Instrum.* **89**, 113707 (2018).

[205] T. Engelhard, "Spinanregungen in FeSe Monolagen", Master's thesis, Karlsruhe Institute of Technology (2017).

[206] T. Frauhammer, "Untersuchung dreibeiniger Tetraphenylmethan-Derivate auf einer Au(111)-Oberfläche mittels Rastertunnelmikroskopie", Master's thesis, Karlsruhe Institute of Technology (2016).

[207] L. Mougel, "Scanning Tunneling Microscopy study of tetraphenylmethane and s-triazine derivatives on Au(111) surface", Master's thesis, Karlsruhe Institute of Technology (2018).

[208] J. V. Barth, H. Brune, G. Ertl, and R. J. Behm, "Scanning tunneling microscopy observations on the reconstructed Au(111) surface: atomic structure, long-range superstructure, rotational domains, and surface defects", *Phys. Rev. B* **42**, 9307–9318 (1990).

[209] D. N. Batchelder and R. O. Simmons, "X-ray lattice constants of crystals by a rotating-camera method: Al, Ar, Au, caf_2, Cu, Ge, Ne, Si", *J. Appl. Phys.* **36**, 2864–2868 (1965).

[210] K. Iwaya, R. Shimizu, T. Hashizume, and T. Hitosugi, "Systematic analyses of vibration noise of a vibration isolation system for high-resolution scanning tunneling microscopes", *Rev. Sci. Instrum.* **82**, 083702 (2011).

[211] M. Liebmann, J. R. Bindel, M. Pezzotta, S. Becker, F. Muckel, T. Johnsen, C. Saunus, C. R. Ast, and M. Morgenstern, "An ultrahigh-vacuum cryostat for simultaneous scanning tunneling microscopy and magneto-transport measurements down to 400 mK", *Rev. Sci. Instrum.* **88**, 123707 (2017).

[212] T. Lutz, C. Große, C. Dette, A. Kabakchiev, F. Schramm, M. Ruben, R. Gutzler, K. Kuhnke, U. Schlickum, and K. Kern, "Molecular orbital gates for plasmon excitation", *Nano Lett.* **13**, 2846–2850 (2013).

[213] P. B. Johnson and R. W. Christy, "Optical constants of the noble metals", *Phys. Rev. B* **6**, 4370–4379 (1972).

[214] R. L. Olmon, B. Slovick, T. W. Johnson, D. Shelton, S.-H. Oh, G. D. Boreman, and M. B. Raschke, "Optical dielectric function of gold", *Phys. Rev. B* **86**, 235147 (2012).

[215] P. R. West, S. Ishii, G. V. Naik, N. K. Emani, V. M. Shalaev, and A. Boltasseva, "Searching for better plasmonic materials", *Laser Photonics Rev.* **4**, 795–808 (2010).

[216] N. Agraït, J. G. Rodrigo, and S. Vieira, "Conductance steps and quantization in atomic-size contacts", *Phys. Rev. B* **47**, 12345–12348 (1993).

[217] F. Silly, A. O. Gusev, E. L. Goff, L. Barbier, and F. Charra, "Correlation between STM-induced photon emission and barrier height: the case of the Cu_3Au alloy vicinal surface", *EPL* **64**, 475–481 (2003).

[218] L. Bartels, S. W. Hla, A. Kühnle, G. Meyer, K.-H. Rieder, and J. R. Manson, "STM observations of a one-dimensional electronic edge state at steps on Cu(111)", *Phys. Rev. B* **67**, 205416 (2003).

[219] D. H. Mosca, F. Petroff, A. Fert, P. A. Schroeder, W. P. Pratt, and R. Laloee, "Oscillatory interlayer coupling and giant magnetoresistance in Co/Cu multilayers", *J. Magn. Magn. Mater.* **94**, L1–L5 (1991).

[220] S. S. P. Parkin, R. Bhadra, and K. P. Roche, "Oscillatory magnetic exchange coupling through thin copper layers", *Phys. Rev. Lett.* **66**, 2152–2155 (1991).

[221] O. Pietzsch, S. Okatov, A. Kubetzka, M. Bode, S. Heinze, A. Lichtenstein, and R. Wiesendanger, "Spin-resolved electronic structure of nanoscale cobalt islands on Cu(111)", *Phys. Rev. Lett.* **96**, 237203 (2006).

[222] J. J. de Miguel, A. Cebollada, J. M. Gallego, R. Miranda, C. M. Schneider, P. Schuster, and J. Kirschner, "Influence of the growth conditions on the magnetic properties of fcc cobalt films: from monolayers to superlattices", *J. Magn. Magn. Mater.* **93**, 1–9 (1991).

[223] A. Cebollada, R. Miranda, C. Schneider, P. Schuster, and J. Kirschner, "Experimental evidence of an oscillatory magnetic coupling in Co / Cu / Co epitaxial layers", *J. Magn. Magn. Mater.* **102**, 25–29 (1991).

[224] P. Bruno and C. Chappert, "Oscillatory coupling between ferromagnetic layers separated by a nonmagnetic metal spacer", *Phys. Rev. Lett.* **67**, 1602–1605 (1991).

[225] M. T. Johnson, S. T. Purcell, N. W. E. McGee, R. Coehoorn, J. aan de Stegge, and W. Hoving, "Structural dependence of the oscillatory exchange interaction across Cu layers", *Phys. Rev. Lett.* **68**, 2688–2691 (1992).

[226] W. F. Egelhoff and M. T. Kief, "Antiferromagnetic coupling in Fe/Cu/Fe and Co/Cu/Co multilayers on Cu(111)", *Phys. Rev. B* **45**, 7795–7804 (1992).

[227] M. T. Johnson, R. Coehoorn, J. J. de Vries, N. W. E. McGee, J. aan de Stegge, and P. J. H. Bloemen, "Orientational dependence of the oscillatory exchange interaction in Co/Cu/Co", *Phys. Rev. Lett.* **69**, 969–972 (1992).

[228] S. S. P. Parkin, R. F. Marks, R. F. C. Farrow, G. R. Harp, Q. H. Lam, and R. J. Savoy, "Giant magnetoresistance and enhanced antiferromagnetic coupling in highly oriented Co/Cu (111) superlattices", *Phys. Rev. B* **46**, 9262–9265 (1992).

[229] A. Rabe, N. Memmel, A. Steltenpohl, and T. Fauster, "Room-temperature instability of Co/Cu(111)", *Phys. Rev. Lett.* **73**, 2728–2731 (1994).

[230] J. de la Figuera, J. E. Prieto, C. Ocal, and R. Miranda, "Scanning-tunneling-microscopy study of the growth of cobalt on Cu(111)", *Phys. Rev. B* **47**, 13043–13046 (1993).

[231] J. de la Figuera, J. E. Prieto, C. Ocal, and R. Miranda, "Surface etching and enhanced diffusion during the early stages of the growth of Co on Cu(111)", *Surf. Sci.* **307-309**, 538–543 (1994).

[232] M. Ø. Pedersen, I. A. Bönicke, E. Lægsgaard, I. Stensgaard, A. Ruban, J. Nørskov, and F. Besenbacher, "Growth of Co on Cu(111): subsurface growth of trilayer Co islands", *Surf. Sci.* **387**, 86–101 (1997).

[233] S. Speller, S. Degroote, J. Dekoster, G. Langouche, J. E. Ortega, and A. Närmann, "Low-temperature deposition of Co on Cu(111): effects on step etching", *Surf. Sci.* **405**, L542–L548 (1998).

[234] J. de la Figuera, J. E. Prieto, G. Kostka, S. Müller, C. Ocal, R. Miranda, and K. Heinz, "Crystallography and morphology of the early stages of the growth of by LEED and STM", *Surf. Sci.* **349**, L139–L145 (1996).

[235] Q. H. Vu and K. Morgenstern, "Upward mass transport and alloying during the growth of Co on Cu(111)", *Phys. Rev. B* **95**, 125423 (2017).

[236] N. N. Negulyaev, V. S. Stepanyuk, P. Bruno, L. Diekhöner, P. Wahl, and K. Kern, "Bilayer growth of nanoscale Co islands on Cu(111)", *Phys. Rev. B* **77**, 125437 (2008).

[237] A. L. V. de Parga, F. J. García-Vidal, and R. Miranda, "Detecting electronic states at stacking faults in magnetic thin films by tunneling spectroscopy", *Phys. Rev. Lett.* **85**, 4365–4368 (2000).

[238] O. Pietzsch, A. Kubetzka, M. Bode, and R. Wiesendanger, "Spin-polarized scanning tunneling spectroscopy of nanoscale cobalt islands on Cu(111)", *Phys. Rev. Lett.* **92**, 057202 (2004).

[239] L. Gómez, C. Slutzky, J. Ferrón, J. de la Figuera, J. Camarero, A. L. V. de Parga, J. J. de Miguel, and R. Miranda, "Novel microscopic mechanism of intermixing during growth on soft metallic substrates", *Phys. Rev. Lett.* **84**, 4397–4400 (2000).

[240] J. Khalil, G. Bozzolo, D. Farías, A. L. V. de Parga, J. J. de Miguel, and R. Miranda, "Atomistic modeling of Co growth on Cu(111)", *Mater. Res. Soc. Symp. Proc.* **696**, N9.3 (2001).

[241] L. Diekhöner, M. A. Schneider, A. N. Baranov, V. S. Stepanyuk, P. Bruno, and K. Kern, "Surface states of cobalt nanoislands on Cu(111)", *Phys. Rev. Lett.* **90**, 236801 (2003).

[242] M. V. Rastei, B. Heinrich, L. Limot, P. A. Ignatiev, V. S. Stepanyuk, P. Bruno, and J. P. Bucher, "Size-dependent surface states of strained cobalt nanoislands on Cu(111)", *Phys. Rev. Lett.* **99**, 246102 (2007).

[243] F. Besenbacher, "Scanning tunnelling microscopy studies of metal surfaces", *Rep. Prog. Phys.* **59**, 1737–1802 (1996).

[244] F. Flores, P. M. Echenique, and R. H. Ritchie, "Energy dissipation processes in scanning tunneling microscopy", *Phys. Rev. B* **34**, 2899–2902 (1986).

[245] L. Bürgi, O. Jeandupeux, H. Brune, and K. Kern, "Probing hot-electron dynamics at surfaces with a cold scanning tunneling microscope", *Phys. Rev. Lett.* **82**, 4516–4519 (1999).

[246] L. Petersen, P. T. Sprunger, P. Hofmann, E. Lægsgaard, B. G. Briner, M. Doering, H.-P. Rust, A. M. Bradshaw, F. Besenbacher, and E. W. Plummer, "Direct imaging of the two-dimensional fermi contour: fourier-transform STM", *Phys. Rev. B* **57**, R6858–R6861 (1998).

[247] O. A. O. Adam, "Detection and manipulation of metallic and magnetic nanostructures: an STM study on (sub)surface atoms, cavities and islands", PhD thesis, Technische Universiteit Eindhoven (2008).

[248] W. L. Barnes, "Surface plasmon–polariton length scales: a route to sub-wavelength optics", *J. Opt. A: Pure Appl. Opt.* **8**, S87–S93 (2006).

[249] Y. Uehara, Y. Kimura, S. Ushioda, and K. Takeuchi, "Theory of visible light emission from scanning tunneling microscope", *Jpn. J. Appl. Phys.* **31**, 2465–2469 (1992).

Acknowledgments

Building this machine and actually getting it to work was a *huge* project that required a lot of innovative problem solving, equipment and expertise from other groups, as well as raw manpower. It would have been entirely impossible to get this done all by myself. I would like to express my deep gratitude and thank the following people in (almost) no particular order:

- Prof. Dr. Wulf Wulfhekel for providing me the unique opportunity to build my own machine from scratch, trusting me with this as if it was self-evident, and supporting me throughout my work with valuable practical suggestions and physical insights. It is truly admirable how much he cares for everyone in the group.
- Prof. Dr. David Hunger for co-examining this work with delightful enthusiasm for the topic.
- Dr. Lukas Gerhard for constantly pushing me to keep things simple, his persistent and unmatched tendency to think outside of the box, and for rigorously questioning absolutely everything. Working with him was a true pleasure!
- Moritz Winkler, Lars Wilmes and Vibhuti Rai who worked closely with me during different stages of the project. They contributed important ideas, lightened my workload, forgave me my stubborn perfectionism, and always provided a pleasant and joyful working atmosphere.
- Prof. Dr. Martin Wegener, Dr. Martin Schumann and Dr. Christian Kern for providing the DLW mirror tips. Without their commitment, this project would have never turned out so nice.
- Simone Dehm for performing many SEM measurements and providing me custom nano-structured samples for the Praktikum.
- Dr. Arne Fischer and Dr. Thomas Reisinger for cutting the metal sheets and ceramic parts with their nice laser.

- The colleagues from the chemistry department Hagen Sparrenberger, Dr. Marcin Lindner, Dr. Jasmin Santoro and Andreas Przybilla for occasionally providing tasty solvents and acids, and handling them for me in the cases when I did not dare to do it myself.

- Dr. Arnold Seiler for calibrating the thermometer and for letting me use their gold evaporation setup for an entire week.

- Dr. Sergei Lebedkin for *a lot* of invaluable support regarding the spectrometer and the numerical aperture matching.

- Prof. Dr. Ralph Krupke for technical advice on my optical setup and highly interesting discussions.

- Prof. apl. Dr. Detlef Beckmann, Dr. Cahit Benel, Dr. Ralf Witte and Dr. Christian Reitz for lending me UHV and other hardware whenever needed.

- Prof. Dr. Marcel Mayor, Dr. Michal Valášek, Prof. Dr. Luisa De Cola, Dr. Rémi Rouquette and Prof. Dr. Birgit Weber for providing molecules for our studies, even though they did not end up in my thesis.

- Dr. Timofey Balashov for quite some important discussions about STM design in general and how to actually do things in particular.

- The other current and former group members Timo Frauhammer, Jan Homberg, Loïc Mougel, Dr. Jasmin Jandke, Dr. Marie Hervé, Dr. Jinjie Chen, Dr. Hironari Isshiki, Dr. Fang Yang, Clemens Baretzky and Tobias Engelhard. They were always helpful when I needed to borrow equipment, and we worked on joint projects, had many valuable discussions and unforgettable social events.

- The workshop teams of Michael Meyer at PHI and Michael Schlenker at INT for manufacturing many — partially mind-blowingly complicated — parts for me.

- Roland Jehle and Ulrich Opfer for providing custom electronics and repairing broken devices.

- The INT infrastructure team Jürgen Harich, Sven Stahl, Sven Helfrich, Albert Schmitt, Andreas Neumann, Martin Limbach, Thomas Koch, Marc Schleifer, Michael Suck and Dr. Matthias Hettler for all sorts of practical support in the lab, office and workshop. And, of course, the traditional coffee breaks.

- The management and administrative staff of both the INT and PHI, in particular Dr. Brigitte Baretzky, Susanne Speck, Antje Hase, Christine Fischer, Patricia Jäger, Heidi Hagel, Steffi Baatz, Ilona Brosch and Birgit Schelske for being incredibly nice, helpful and quick whenever I needed anything.

- My entire family for supporting me throughout this crazy endeavour, especially when it came to cheering me up when things went wrong (which they absolutely did), taking care for my children when I couldn't, or maintaining my car, without which I would not have been able to work at all.

Experimental Condensed Matter Physics
(ISSN 2191-9925)

Herausgeber
Physikalisches Institut

Prof. Dr. David Hunger
Prof. Dr. Alexey Ustinov
Prof. Dr. Georg Weiß
Prof. Dr. Wolfgang Wernsdorfer
Prof. Dr. Wulf Wulfhekel

Band 1 Alexey Feofanov
Experiments on flux qubits with pi-shifters. 2011
ISBN 978-3-86644-644-1

Band 2 Stefan Schmaus
Spintronics with individual metal-organic molecules. 2011
ISBN 978-3-86644-649-6

Band 3 Marc Müller
Elektrischer Leitwert von magnetostriktiven Dy-Nanokontakten. 2011
ISBN 978-3-86644-726-4

Band 4 Torben Peichl
Einfluss mechanischer Deformation auf atomare Tunnelsysteme – untersucht mit Josephson Phasen-Qubits. 2012
ISBN 978-3-86644-837-7

Band 5 Dominik Stöffler
Herstellung dünner metallischer Brücken durch Elektromigration und Charakterisierung mit Rastersondentechniken. 2012
ISBN 978-3-86644-843-8

Band 6 Tihomir Tomanic
Untersuchung des elektronischen Oberflächenzustands von Ag-Inseln auf supraleitendem Niob (110). 2012
ISBN 978-3-86644-898-8

Band 7 Lukas Gerhard
Magnetoelectric coupling at metal surfaces. 2013
ISBN 978-3-7315-0063-6

Band 8 Nicht erschienen.

Die Bände sind unter www.ksp.kit.edu als PDF frei verfügbar oder als Druckausgabe bestellbar.

Band 9 Jochen Zimmer
Cooper pair transport in arrays of Josephson junctions. 2014
ISBN 978-3-7315-0130-5

Band 10 Oliver Berg
Elektrischer Transport durch Nanokontakte von Selten-Erd-Metallen. 2014
ISBN 978-3-7315-0209-8

Band 11 Grigorij Jur'evic Grabovskij
Investigation of coherent microscopic defects inside the tunneling barrier of a Josephson junction. 2014
ISBN 978-3-7315-0210-4

Band 12 Cornelius Thiele
STM Characterization of Phenylene-Ethynylene Oligomers on Au(111) and their Integration into Carbon Nanotube Nanogaps. 2014
ISBN 978-3-7315-0235-7

Band 13 Michael Peter Schackert
Scanning Tunneling Spectroscopy on Electron-Boson Interactions in Superconductors. 2014
ISBN 978-3-7315-0238-8

Band 14 Susanne Butz
One-Dimensional Tunable Josephson Metamaterials. 2014
ISBN 978-3-7315-0271-5

Band 15 Philipp Jung
Nonlinear Effects in Superconducting Quantum Interference Meta-Atoms. 2014
ISBN 978-3-7315-0294-4

Band 16 Sebastian Probst
Hybrid quantum system based on rare earth doped crystals. 2016
ISBN 978-3-7315-0345-3

Band 17 Wolfram Kittler
Magnetische Anisotropie und Quantenphasenübergang in $CeTi_{1-x}V_xGe_3$. 2015
ISBN 978-3-7315-0363-7

Band 18 Moritz Peter
Towards magnetic resonance in scanning tunneling microscopy using heterodyne detection. 2015
ISBN 978-3-7315-0410-8

Die Bände sind unter www.ksp.kit.edu als PDF frei verfügbar oder als Druckausgabe bestellbar.

Band 19 Junji Tobias Märkl
Investigation of Magnetic Adatoms with Scanning Tunneling Techniques. 2015
ISBN 978-3-7315-0435-1

Band 20 Jochen Braumüller
Quantum simulation experiments with superconducting circuits. 2018
ISBN 978-3-7315-0780-2

Band 21 Jinjie Chen
Local Investigation of Single Magnetic Molecules with Scanning Tunneling Microscopy. 2018
ISBN 978-3-7315-0819-9

Band 22 Arnold Seiler
Einfluss der Leitungselektronen auf die Dynamik atomarer Tunnelsysteme in ungeordneten Festkörpern: Relaxationsprozesse in metallischen Gläsern und ungeordneten dünnen Aluminiumoxid-Schichten. 2019
ISBN 978-3-7315-0870-0

Band 23 Jasmin Maria Jandke
Elastic and Inelastic Scanning Tunneling Spectroscopy on Iron-Based Superconductors. 2019
ISBN 978-3-7315-0747-5

Band 24 Kevin Edelmann
Electroluminescence from Plasmonic Excitations in a Scanning Tunnelling Microscope. 2019
ISBN 978-3-7315-0923-3

Die Bände sind unter www.ksp.kit.edu als PDF frei verfügbar oder als Druckausgabe bestellbar.